MW00582249

HILDE ØSTBY

Translated by Matt Bagguley

THE Key to Creativity

The Science Behind Ideas and How Daydreaming Can Change the World

GREYSTONE BOOKS
Vancouver/Berkeley/London

First published in English by Greystone Books in 2023
Originally published in Norwegian as *Kreativitet*,

Illustrations by John Tenniel

23 24 25 26 27 5 4 3 2 1

Greystone Books Ltd.
greystonebooks.com

Cataloguing data available from Library and Archives Canada
ISBN 978-1-77164-830-1 (cloth)
ISBN 978-1-77164-831-8 (epub)

Editing for English edition by James Penco
Proofreading for English edition by Meg Yamamoto
Jacket and text design by Fiona Siu
Jacket artwork by John Tenniel and Edvard Munch
Printed and bound in Canada on FSC® certified paper at Friesens.
The FSC® label means that materials used for the product have been responsibly sourced.

Greystone Books thanks the Canada Council for the Arts, the British Columbia Arts Council, the Province of British Columbia through the Book Publishing Tax Credit, and the Government of Canada for supporting our publishing activities.

British Columbia Arts Council
An agency of the Province of British Columbia

Greystone Books gratefully acknowledges the xʷməθkʷəy̓əm (Musqueam), Sḵwx̱wú7mesh (Squamish), and səl̓ílwətaʔɬ (Tsleil-Waututh) peoples on whose land our Vancouver head office is located.

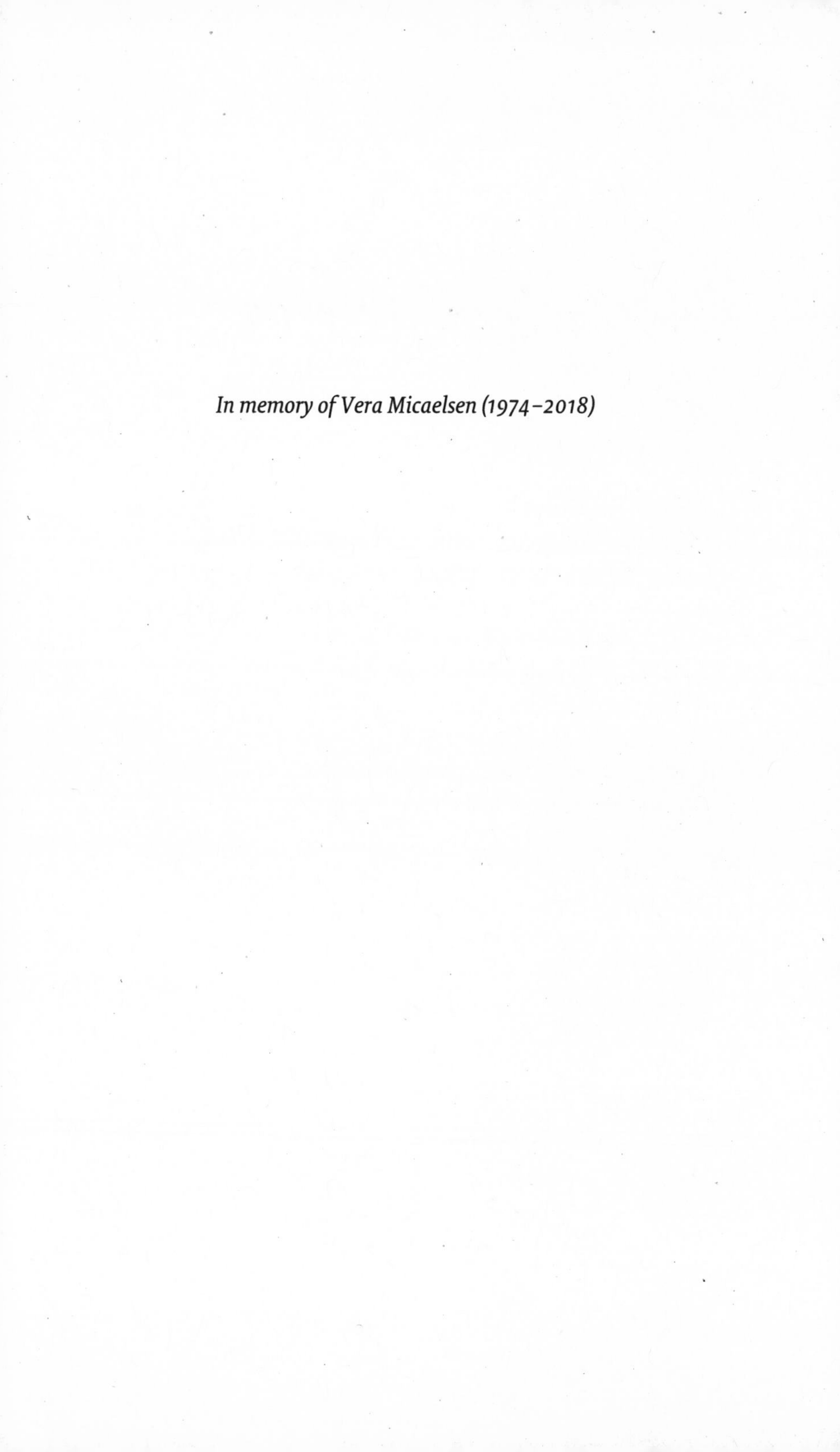

In memory of Vera Micaelsen (1974–2018)

Contents

"She had never before seen a rabbit with either a waistcoat-pocket, or a watch to take out of it, and burning with curiosity, she ran across the field after it, and fortunately was just in time to see it pop down a large rabbit-hole under the hedge."

LEWIS CARROLL,
Alice's Adventures in Wonderland

INTRODUCTION

I Hit the Wall by the River Akerselva

OR: CAN BUMPING YOUR HEAD MAKE YOU MORE CREATIVE?

SO I HIT the wall. Literally.

It was a stone wall, which I could almost taste when I smashed into it. It tasted like... stone. A cold, strangely metallic taste. Or that could have been just the taste of blood.

It was the day my sister Ylva and I were due to launch a book we had written together called *Adventures in Memory*; we were scheduled to meet radio and press journalists to talk about the book, which was all about memory and the brain. I had just delivered—well, more like thrown—my daughter at kindergarten and was cycling hastily along the riverside path to work, following the unbroken strip of steel-blue water that cuts through the middle of Oslo, under bridges and past long embankments of gray autumn grass. My heart pounded as I mentally prepared myself for what I was supposed to be doing that day. Then—in a split second of distractedness—I turned my head, convinced there were a good few feet remaining between my bike and the low bridge arching over the path ahead of me.

When I turned around again, there it was—the bridge—stout and steadfast as it had been since 1827. It didn't move an inch as I slammed into it, although I think it should have, out of pure courtesy; we're talking about a grand old gentleman, raised in the early 1800s—and this was definitely no way to treat a lady. When I finally hit the ground after what felt like minutes—time slowed down, it really did—it was with quite a thud. My bike, which had just trundled on as my head struck the masonry, now lay several feet away.

My face was a pulverized mess of cuts and bruises, and a giant lump had sprouted from my forehead. Blood cascaded down my brown overcoat, and I found out later that my nose was broken. Ironically, and luckily for me, the bridge I'd just collided with was right next to the local Emergency Room.

The bridge, called Nybrua, was once a proud new addition to the city's road network and an unquestionable boost to the lives of those living in Christiania, as Oslo was called at the time. Now, as I staggered over it with help from a passing jogger, it was both my worst enemy and my savior. Moments later I lurched through the doors of the ER, massive forehead first, where they sent me for a CT scan to check for intracranial hemorrhaging.

I can safely say that my life has been turned upside down because of what happened to my head beside the river that day. That ordinary Tuesday in late October.

But what had actually happened?

After a couple of days, I'd developed what's called "periorbital ecchymosis," dark blue rings around my eyes, like a raccoon. But by then I'd already been sent home from the ER with a leaflet explaining that I shouldn't do any reading (that's right, I read a leaflet about not reading and felt somewhat tricked), that I should take it easy for three weeks, and that everything would be okay. I had a mild concussion, apparently. It wasn't dangerous. I hadn't even fainted. I'd just had a little knock. Thousands of people are similarly injured every year. It's so normal that it's almost not worth trying to evoke sympathy from a reader.

So I was now just a statistic—a "cyclist without a helmet with a head injury." A TBI (Traumatic Brain Injury), according to medical literature. According to the Head Injury Severity Scale (HISS), which has been used in Scandinavia since 1995, I didn't even have a *mild* head injury. Since I hadn't fainted, and therefore hadn't lost consciousness, my injury was classified as a "minimal head injury." So, theoretically, the severity of my accident was so minuscule I was unlikely to experience any symptoms—and would soon be totally healthy again. But even while I'd been writhing around in the dirt, bleeding and in shock, I'd already suspected that what had happened might lead to memory problems. It wouldn't be unusual, I thought, since memory is so fragile and involves so many networks throughout the brain. After all, I have a sister with a PhD in neuropsychology and memory—and, as I already mentioned, I had just cowritten a book with her all about the subject.

But it turned out, unsurprisingly, that my self-diagnosis wasn't entirely correct. After the accident, my memory was still crystal clear—not just regarding my life experiences, but also on all the scientific research I'd read while writing the memory book. My brain seemed to be injured in some other way, and besides—was it really *injured*?

Surprisingly enough, right after the crash I found myself with an incredible number of ideas, and bursting with energy. Lying still felt totally impossible. In the week following the collision, I sat down to write a short list of ideas for potential nonfiction books—and instead of a few ideas, I came up with twenty. Had the blow to my head made me more creative and more efficient than normal? When I talked to my doctor about this, she laughed and said, "Perhaps everyone should get an occasional bump on the head?" I laughed too, joking that I would take out a patent for "Hilde's Blunt Force Method." Little did I know that it would be the last joke I would crack for a very long time.

Research shows that some people have experienced huge bursts of creativity following a blow to the head or trauma, be it mental or physical, because it has been shown to destabilize the levels of dopamine in the brain. Dopamine is the brain's reward system and, when it comes to creative expression, it can make your head buzz with ideas. More specifically, I had read that you can become more creative from an injury to the temporal lobe.

"Mania and depression can come in complicated mixtures," writes the neurologist Alice Weaver Flaherty, who became hypergraphic after the premature birth of twin boys who died after only a few days—a huge trauma for a young mother. Afterward, she had an irrepressible urge to write, regularly getting up in the middle of the night to sit in front of the computer. She believes that the temporal lobe is the key to creativity; it is here that both hypergraphia and writer's block originate.

"Even now, when I am writing well, my pulse speeds up, I feel gripped by something stronger than my will, and I have some of the delicious feeling I had at my most hypergraphic."

After reading Flaherty's story, I started reading far more sobering research about blows to the head, including an article about the renowned French composer Maurice Ravel. Those researching his case believe that a car accident was probably to blame for him losing the ability to write music. There are far more examples of people becoming *less* creative, not more, after a head injury or trauma. In my case, it was impossible to know for sure—perhaps my brain had entered a manically creative artist mode?

Much later—after many months of being unable to work—my doctor and I were no longer laughing about my Blunt Force Method. I had suffered a number of strange breakdowns where I cried uncontrollably, like a child—like my own child in fact (small children will cry frequently and apparently for no reason)—because there was too much noise or commotion. Any kind of sound made me feel completely exhausted. When walking through Oslo Airport's duty-free area I had to curl up on a bench, gasping for breath, in a fetal position. The sensory impressions around me were just too much. I was sleeping at least twelve hours a night, yet still wanted an afternoon nap because I was so tired.

My working memory—my executive function—had clearly been affected.

Executive function is very important for what we call focus and concentration. Like the captain of a ship, executive function guides your thoughts. Whether you are calculating something in your head, replying to an email, or sitting in a meeting, you need your executive function in order to be focused, alert, and able to manage several thoughts at once. Memory researchers believe that between five and nine "units"—such as numbers or items on a shopping list—can be moved around in your working memory simultaneously. With a working executive function, you can also plan and gauge the consequences of what you are doing, create future scenarios, and retrieve memories. But when my captain was knocked out, each impression I experienced came at me with equal intensity, and the future became a whirl of confusion.

Executive function is not just the captain either; it is the ticket inspector, and without a ticket inspector, sensory impressions can just come on board, quarrel with your brain, and then fall over the side as the ship sails away, with nobody at the helm and the whole voyage descending into chaos.

Much later, I found a scientific article describing how traumatic brain injury (TBI) can affect the brain's different networks—both executive function, which I have just described, and what is called the default mode, or "daydream," network (DMN)—and how they interact. At the time I knew very little about what this meant, but I was convinced that something had been inhibiting me in the past that was not inhibiting me now, after my accident. My ideas and daydreams were now insistent, wild, and unstoppable.

I already knew that the brain is wildly associative, and that we are, at any given time, controlled by strong feelings, memories, and associations. And since I knew that memory is highly creative and unpredictable, I believed existing memory research could explain what had happened to me and my creativity. Countless experiments have shown that we are able to remember things that never happened, and we probably remember many of the most important events in our lives incorrectly. Memory is a faulty tape recorder at its worst, and a teller of fairy tales at best. The first law of memory is that we remember everything that stands out or evokes a strong emotion; trauma occupies far more space in our memory than brushing our teeth. And since most of us have a limited capacity for memories, the second law applies—we bundle normal and mundane events, like brushing our teeth, into so-called collective memories. But even here your brain's creativity is already at work; the things you forget and the things you remember are governed by your feelings, by what you personally think is important. In addition to this, our memories are shaped and altered whenever they are retrieved.

Nevertheless, memory research cannot describe every aspect of creativity, because the brain can produce strange, spontaneous ideas and "aha" moments that seem completely detached from both the past and the present—thoughts like "what if?" which after my accident came to me frequently. This is something memory researchers know very little about. So what do we actually *know* about our most amazing and marvelous attribute—our capacity for creative thinking?

Creative thinking has given us palaces and pyramids, the moon landings and the *Mona Lisa*, waterwheels and automobiles, the most incredible discoveries, fantastic cities, and amazing technological solutions that have allowed us to conquer the planet. Our creativity means that scientists now talk about the "Anthropocene," an era where humankind, not nature, is leaving huge, everlasting scars on the planet. Human creativity is deeply and profoundly connected to our position on earth, and for the last two hundred years we have used this power to such a great extent that it may lead to catastrophe. The climate crisis is one of the most direct consequences of our shared creative ability to survive, our surfeit of energy, our curiosity, knowledge, and the stories we tell each other.

While I lay on the sofa, trying to get my concussed and overactive head to calm down, I thought about *Alice's Adventures in Wonderland*, a book we had recently discussed in the reading club my friend Vera and I had established many years earlier. Vera loved knitting, and I thought about the way her knitting needles had moved while we discussed the book. I tried to remember what she had said about the White Rabbit; it felt important. It felt like I was mentally chasing a stressed white rabbit as I lay there, without really understanding why—just like Alice.

Some people help us change the way we think, offering us vivid and colorful images and dreams, simply because their inner world is so wild and urgent that it needs to be shared. Lewis Carroll's *Alice's Adventures in Wonderland* left an indelible mark on modern culture, and when I read it again as an adult I understood how radical the book is—it taught me that stuff and nonsense is highly important and should be taken with the utmost seriousness. I also felt like I was finding myself in Wonderland quite regularly now, after my knock on the head.

Carroll's legendary children's book was conceived on a beautiful July day while boating on a lazy river in Oxford. The book you have in your hands now was born from a trip along an equally lazy river on a cloudy October day in Oslo more than 150 years later. Since that day, much of what I once took for granted has vanished into thin air, like the Cheshire Cat leaving with nothing but a smile. Now I see life quite differently. I moved house, quit my job, and—for the first time in my life—I now earn a living from my creativity, which could be precisely why I need to find out what creativity is and how it works. What I have realized is that anything can happen, and usually it does, when you least expect it. Because what happens if you do follow a white rabbit down a hole one warm summer day? What happens if you crash your bike into a bridge one cold autumn day, and your life is never the same again?

What if this event triggered something quite unexpected? What would happen if I wrote a whole book about creativity?

We are about to find out.

1 | The Cheshire Cat Appears

OR: GOOD IDEAS, AND EIGHT HUNDRED "AHA" MOMENTS THAT CAN TURN YOUR LIFE UPSIDE DOWN.

"When I was your age, I always did it for half-an-hour a day. Why, sometimes I've believed as many as six impossible things before breakfast."

ONE OF THE most influential ideas of the modern age popped into the head of a twelve-year-old in 1900, in a tiny little Scottish village called Helensburgh.

This idea, which must have sounded like pure nonsense to all those around the young boy, would completely change modern life—and is now taken for granted all over the world. It is quite possible that you have never heard of him, but you will certainly have heard of his invention; the average American spends three hours of their day on what John Logie Baird invented in between the bouts of constipation, influenza, and bronchitis that plagued him throughout his childhood. And this crazy idea—the television—pursued him for his entire life.

So where did it come from? Baird was a smart and technically gifted twelve-year-old who, for example, managed to light his entire house with homemade electricity—generated from a waterwheel, a dynamo, and some lead sheets wrapped in flannel that he'd immersed in sulfuric acid. At the same time, this little boy began studying the village telephone system, and built a copy of it at home. And it was while he was making his home telephone system that he had the groundbreaking idea that would write him into the history books: What if a telephone sent not only sound, but pictures? Baird called the idea "seeing by wireless."

After completing his engineering studies, he began displaying a relentless level of get-up-and-go. For a while he ran a mango chutney factory in Trinidad, where he boiled the ingredients in an outdoor bathtub—into which every nearby insect fell and drowned, while cockroaches swarmed around the sugar bags. Later on, he came up with the idea for a kind of air-cushioned shoe—a boot he had lined with balloons. He also attempted to make a rust-free razor blade out of glass, but his dream was duly crushed when he cut himself badly while shaving with the prototype. But even when he made it big selling soap and as the inventor of the heated sock, it still wasn't enough for him. He had a dream, and he refused to give up before trying to make it happen.

In 1924, Baird assembled the first model of his seeing telephone—using a tea chest, a hatbox, a projection lamp, a lens from a bicycle light, glue, string, and some high-voltage wire—and gave himself a two-thousand-volt shock in the process. In 1925, he met with a representative of the Italian company Marconi in London and asked if they were interested in a partnership. But he was told they were absolutely not interested in the so-called "television." Later Baird described the rejection as being like he had asked them to invest in a bordello.

"This episode shows the general attitude to television in 1925. It was regarded as a wildcat myth, something on a par with the Perpetual Motion Machine. Television could never be realised unless some hitherto undreamt of discoveries were made, and nothing of the sort was in sight," he wrote in his entertaining memoir *Television and Me*.

There was, however, one thing that would pave the way for success. Thanks to the German engineer Paul Gottlieb Nipkow, the Scot now had the most important component for producing his television: a rotating disk containing lenses called a Nipkow disk, which divided an image into dozens of flashing lines. While Baird worked on his own homemade set, he was often in danger of being struck by the lenses—which would break free of the rotating disk and fly through the air before smashing against the wall. Anyone else would have given up then and there. But Baird was so obsessed with his vision that none of these dangers worried him.

From the moment he initially got his idea in Helensburgh, twenty-five years would pass before it was realized in 1925. Unbelievably, it was in the high-end department store Selfridges on Oxford Street—London's fashionable West End—that John Logie Baird, the former soap seller from a tiny village in Scotland, demonstrated his new invention. It was here that the world's very first television images were seen.

All this may seem momentous today, but it wasn't a particularly exciting broadcast. The images consisted of white figures on a black background, and a high contrast was required for them to be recognizable, since the resolution was so poor. But the miracle had happened: Baird had succeeded in sending an image from one place to another, and had laid the foundation for the concept of modern-day TV broadcasting.

"A potential social menace of the first magnitude," exclaimed Sir John Reith in describing the invention. Reith was the first general manager of the BBC, and compared the TV set to smallpox and the Black Death. At the time, the British Broadcasting

Corporation was a company specializing in radio broadcasts, and it would take years before it changed its view of Baird's discovery.

For Nipkow, who had dreamed of making a TV long before Baird but lacked the technical conditions to see it through, it must have been strange to stand in a queue in Berlin in 1928, waiting to experience the newest miracle everyone was talking about. At the end of the queue, he would have seen the pictures flickering across Baird's apparatus, on display for curious Germans, keen to watch TV for the first time. Forty-five years had passed since Nipkow had patented his disk and dreamt of creating living pictures on a screen. Like Baird, he had sacrificed a great deal, but unlike Baird, it hadn't paid off for him.

What we can already see here is that even if someone has a fantastic idea, it doesn't necessarily lead to success. Most ideas have to be somehow connected to the world around them; they have to resonate and be realized. The people around the idea need to believe in it, understand it, help to make it happen, and spread it. If someone wants to succeed with a good idea, they need to be in the right place at the right time.

When Arnfinn Hegg had his good idea, for example, nobody wanted to touch it—which is why he isn't a billionaire today.

"I was just out for a walk. I'm not sure I was even thinking about skiing at all; there wasn't any snow," the inventor tells me.

For forty-one years, Hegg worked as a dentist, but since boyhood he had gone around looking for problems—well, *solutions* to problems, mostly—that nobody had yet discovered.

"Even in the 1970s I was thinking: How can we put men on the moon and yet have cross-country skis that slide backward? It didn't add up, the level of technology. There had to be some way of fixing these non-grip skis!"

Once he'd discovered the problem, it haunted him. For a long time. And then, one warm August day in 1992, the solution came to him, when he least expected—twenty years after he had first pondered it.

"I was tying my walking shoes before setting off on a hike, and I suddenly realized that there had to be a difference between the level of a ski's grip-zone and slip-zone," he says, referring to the day the idea struck him: a new solution to an age-old skiing problem.

It was then, in the summer of 1992, that "Fantaski" was born. Several visits to ski manufacturers across Norway were made to acquire some knowledge about how modern skis are built—followed by a lot of trial and error, and many prototypes, built at home in Hegg's basement. The solution, in the end, was a ski equipped with a felt strip that prevented the skis from sliding backward. To bring the strip into contact with the surface, the skier just had to tilt the ski a little.

"I haven't used the herringbone technique for twenty-five years. I just ski straight up!" says Hegg, satisfied.

But he wasn't a part of the skiing world, nor was he involved in ski production. So how could he make these skis readily available?

"I was in touch with the technical manager at one of Norway's biggest ski manufacturers, and he was keen to sign a contract with me. But the board members were afraid of committing to the idea," Hegg explains.

He's not bitter. Just slightly amazed. Because now, twenty-five years later, all the big ski producers make "skin skis." Not identical to his patented solution, but using the same principle: the felt attached to the bottom of the skis makes them grip the snow and stops them from sliding backward.

Centuries earlier, in 1480, Leonardo da Vinci experienced something similar: he invented a kind of helicopter, although his invention wouldn't see the light of day for another 420 years. The technological conditions—or any appreciation—for what he'd invented didn't exist back then. Nobody understood what a helicopter would mean for transport and air travel, and the idea was shelved until the time was right.

Inventors were perhaps most in vogue in the eighteenth and early nineteenth centuries, while city populations thrived, technology evolved, and factories and trains first facilitated the mass production and transport of goods; markets and trends were given far more consideration than before. Many people dreamed of making it big as an inventor, many of them willing to sacrifice a great deal to nudge humanity yet another inch toward comfort and modern civilization. One of my favorite books is called *Inventions That Didn't Change the World,* which is about patents that never really caught on: A cigar holder you could attach to your hat, for those who always wanted a cigar at the ready. A highly advanced top hat that could be turned easily into a bowler hat. A shoe with a rotating heel, so you could turn the heel around when a part of it had worn down. We can laugh at these things now, but for the inventors it was, of course, difficult to know—perhaps this would be the next big thing.

Many inventors have also died from their own inventions. One of those who sacrificed everything for his idea was Franz Reichelt.

In a film clip from 1912, shot on a cold February day in Paris, you can actually see him yourself on the internet, looking down at the ground far below. You can tell how cold it is from the cloud of breath rising from his mouth, and you can imagine how hard and fast his heart must have been pounding in his throat, his eardrums, and his fingertips. He probably hadn't noticed himself, because he was too focused on what he was about to do. The film shows him wrapped in a large piece of fabric, molded peculiarly around his body, ready for his terrifying stunt. Since he had told everyone what he was going to do, and had even brought two cameramen along, it was perhaps difficult to back out. He was thirty-three years old at the time, and had created a fair amount of hype as the "Flying Tailor." This was to be the day he would achieve his definitive breakthrough.

His idea had been to make a parachute that could be released if a plane malfunctioned in the air—something that regularly happened in the early days of aviation. There's no doubt that it was a good idea! Modern-day fighter pilots can eject with a parachute on their backs if their plane is about to crash, just as Reichelt had imagined. The problem with the Flying Tailor's invention was that it had only ever been tested at home in Reichelt's apartment, using dummies dropped from the height of the ceiling. When he applied to the Paris authorities for permission to use the Eiffel Tower as a launch site, he had specified that a dummy would be used, not a person. Was he already aware that he was instead going to test it himself? Or did the idea come to him later? Did he understand that he was risking his life, or was he that confident about his own invention?

The cameraman waiting below would have gotten only a brief glimpse of Reichelt as he hurtled full speed toward the ground. The parachute never opened.

Afterward, the film clip, showing Reichelt falling headfirst from the Eiffel Tower, was shown to the horrified public. He had to be almost scraped out of a small crater before he could be buried.

When I look at this video, I wonder how far I would be willing to go for something I thought was a good idea. Would I sacrifice my life? When Franz Reichelt stood there, looking down from the edge of the Eiffel Tower, his breath floating in the cold air and his heart pounding in his ears, surely he would have questioned for a moment whether his invention was worth dying for. Or does a good idea render you deaf and blind to doubters and naysayers? Why did he throw himself toward the ground from 187 feet up; who *does* something like that? Was the Flying Tailor just stark raving mad?

• • • •

WHEN A PERSON makes a great discovery, they might lose touch with rules and conventions a little; it's perhaps even necessary for getting a good idea. They must feel such a strong sense of insight and enlightenment—a sense of knowing something so important that their other emotions, like fear, get pushed aside. When the Greek mathematician Archimedes had one of his best ideas, for example, he no longer cared about the fact that he was naked in public.

When I was in Sicily a few years ago, I visited the place where one of the world's most famous "aha" moments happened. On the southwestern tip of this island, west of the Italian mainland—yes, the football balanced on the tip of the Italian boot—I rented an apartment with a roof terrace in the small city of Syracuse. The city was founded by the Greeks and had been the island's capital thousands of years earlier; today it contains the ruins of 2,500-year-old temples and a Greek theater where operas are still performed on warm summer nights. During antiquity, this was a grand and important city, a place the influential philosopher Plato visited many times. Now it is just a little dot on the Italian map, far from the center of power.

After a week of stiflingly hot days in Syracuse, where the temperature hit 104 degrees Fahrenheit in the shade, I suddenly understood why someone might want to run naked through the city. Because, as I said, that's precisely what Archimedes did, 2,200 years ago.

Hiero, who was the king of Syracuse at the time, had recently had a new crown forged, but was suspicious that the goldsmith may have stolen some of the gold provided and replaced it with silver. To prove this, he could have remelted part of the crown, but that would have destroyed it, of course—and it would have been highly annoying to find out that it was pure gold after all. Hiero turned to his trusted man Archimedes, who was the greatest mathematician of his time and today is viewed as one of the most important mathematicians ever. Among other things, he derived the number pi and created a formula for calculating the area of a circle and volume of a sphere, things we learn about at school to this day. He also invented several machines for defending the city, which, as mentioned earlier, was important back then and of great interest to the king's greedy enemies.

Archimedes thought about the king's problem and went home to take a bath. While sitting in the bathtub, he suddenly realized that the water he displaced was equal to the weight and volume of his body—and that material with a high density, like gold, would displace less water than material with a lighter density, such as human beings—or silver. In an instant he had solved the king's problem. We can picture Archimedes now, gleefully realizing this as he lowered himself into the water. His discovery was so groundbreaking for him that he left the bath immediately and ran dripping wet and stark naked through the streets of Syracuse while shouting, "Eureka! Eureka!" I could almost hear the echoes of his jubilant cries while I sweated my way around the city. And yes, the gold crown really had been mixed with silver, and the thief was duly punished.

So this is, of course, why the Greeks believed that their ideas were sent to them from the gods—because an idea will simply appear, like the Cheshire Cat, and it will surprise you, smiling from the branch of a tree, when you least expect it. Even today, a good idea that just pops into your head can feel almost magical. In ancient Greece, they believed a person could get ideas through a form of divine madness, which at the time was divided into four types—none of which should be confused with today's acute psychological disorders, the types that render us unable to participate in society.

The Greeks believed in a *prophetic* madness and a *ritual* madness, which were both connected to religious practice, and an *erotic* madness, which anyone who's been in love can attest to. Then there was the madness brought about by a muse, *poetic* madness, the madness that produces ideas. A person would be inspired and then (luckily) gain insight from the world of the gods. There were nine of these muses, who were the children of Mnemosyne—the goddess of memory—and the supreme god, Zeus. Later, the poet Hesiod would associate each of the nine female figures with their own artistic areas: Calliope was the muse for epic poetry, Clio for history, Erato for love poetry, Euterpe for music, Polyhymnia for hymns, Terpsichore for song and dance, Melpomene for tragic theater, Thalia for comic theater, and Urania for astronomy. But in Archimedes's day, these muses were not "specialized"—they were just nine sisters (sometimes, only three sisters) who gave people divine inspiration. And the mother of these nine muses was memory—the source of all art forms. It is tempting to draw some wisdom from this—that without knowing what has gone before, it is hard to create something new. Without memory, no creativity.

While I was in Syracuse, there was a rare lunar alignment between Saturn and the moon, and on July 26 that summer, they were so close I was able to observe their dance in the heavens from my vantage point on the terrace roof. Had I been alive six hundred years earlier during the Renaissance, I certainly would have believed this alignment was a sign of a new and creative period. In Roman times, the god Saturn was honored at the so-called Saturnalia, a carnival where everything was turned on its head for an entire week: slaves became masters for a short period, then returned to their normal roles. Saturn was the god of carefully orchestrated chaos, frivolity, and exceptions from the rules, and he was celebrated in the middle of winter. But something happened to Saturn during the Renaissance, when the humanist philosopher Marsilio Ficino began talking about the planet at the court of Cosimo de' Medici, the banker who ruled Florence.

Florence at the end of the 1400s could easily be described as the cradle of Renaissance art—its creative capital even. It was where the Medici family gathered the most celebrated artists and intellectuals of their time. Ficino, a central thinker at Cosimo de' Medici's court, spent much of his time translating

Plato from Greek to Latin, to make him more accessible to his current times. Plato describes how knowledge is hidden within us from the moment we are born as lessons from an intangible world, the world of ideas, that we actually know all that we need to know before we are even born, and that learning is about remembering these abstract concepts and ideas again. According to Plato, when a person gains insights about the world, it is like finally emerging from a dark cave and seeing the sun. He was also fascinated by the seemingly magical connections between mathematics and music that the Greek philosopher and mathematician Pythagoras had been able to demonstrate. During the Renaissance, these magical connections were expanded to match the mathematical relationships between planetary orbits, and people spoke excitedly about the "music of the spheres," the music of the universe.

During the Renaissance, people were unwavering in their belief in alchemy and astrology. Even the celebrated astronomers Tycho Brahe and Johannes Kepler offered horoscopes for money. In this transitionary period between the religiously dominated Middle Ages and our modern era, people lived in a semi-magical universe, where the planets and stars and the people on earth invisibly affected each other, and every planet, star, plant, animal, and stone had a secret insight hidden within them. (Some believe this is what made Isaac Newton search, many years later, for the secret and invisible connection between objects: gravity. But that's another story.) That Saturn could genuinely affect the mood of an artist or a man of science did not seem at all unthinkable in the 1400s, and so now Saturn was the planet of the artists, the god of creative power. At the same time, Saturn also gained control over melancholia, which was described by Aristotle as central to creativity: "Why is it that all men who have become outstanding in philosophy, statesmanship, poetry, and the arts are melancholic?" asked the Greek philosopher in the fourth century BCE.

Ficino translated not only Plato, but also the Egyptian mystic Hermes Trismegistus, who was considered a key philosopher during the Renaissance (although forgotten today, for good reason). This meant that Jewish mysticism, number mysticism, and Plato's theory of ideas became combined with Aristotle's thoughts about the creative melancholic; according to Ficino's interpretation, the world was full of secret connections and number magic—artistic inspiration and ideas, along with dark melancholia, were things that came from the planet Saturn. Ficino and the Medici family lived in an age of upheaval, the Renaissance, when thousand-year-old Christian traditions were changing decisively. Suddenly the future lay terrifyingly open, free of the safe grip of religious authority. In a way, they lived during a never-ending carnival, and Saturn became the symbol of lasting change. Artists, scientists, and politicians embarked on shaping a secular world—paradoxically, within a universe they considered to be governed by magical connections and symbols. This behavior, it has to be noted, was pure megalomania, which from then on became a natural part of our culture. Where God and humility had previously been the ideals, a modern world was being created by magicians believing they were demigods, convinced that their ideas came from the planet Saturn—something Ficino believed would also make them deeply melancholic.

Shakespeare was well acquainted with these ideas in 1611 when he wrote *The Tempest*. It was the last play he wrote before his death, about a wizard-artist ruling a tiny island with his daughter Miranda by his side (Miranda is a name Shakespeare came up with, which means *miracle*). By then, Saturn had become the most important planet for artistic inspiration—for political leaders, for scientists and alchemists, and for artists like Shakespeare. So in *The Tempest*, Shakespeare was popularizing an idea that was already well known in the Renaissance. To test his art and magic, Prospero wields control over a small elfish figure, Ariel—a spirit of the air whose name quite literally means "air"—and for an audience in the 1600s, this androgynous fairy was a familiar sight. Alchemists talked about a "fifth element," the mysterious metal quicksilver, which became associated with Hermes, the flying god of communication who ruled over all that was "in between." This element was crucial to the alchemists, who wanted to have some control over life and death. Artistic inspiration and scientific insight drew from many sources during the Renaissance, and could come to the recipients from Hermes, Ariel, or a muse, or directly from Saturn. Shakespeare helped these ideas take root in the public consciousness, and to this day the concept of inspiration is something we take for granted. The word is derived from the Latin *inspira,* which means "to breathe in"—it is something that comes to us from the surrounding air, like a winged spirit filling us with insight. And in a way we still think like that; ideas come suddenly, almost magically to us, energizing us, like an open window allowing fresh air into the room.

But since we don't believe in air spirits anymore, where do ideas really come from?

What about "aha" moments? In neuropsychology, there is actually a field of research dedicated to them. To a degree, it is possible to explain why Archimedes happily ran naked through Syracuse. Rolf Reber is professor of cognitive psychology at the University of Oslo, and he and his research team have compiled a database of eight hundred "aha" moments. Now he is busy analyzing the "aha" moment's basic components.

"The brain processes things that we recognize much faster, and we find that comforting. Thinking quickly is comfortable—thinking slowly, not so comfortable. According to research, this need to process things fast increases when we feel insecure. Insecurity makes us more fond of repetition, because it makes everything flow more smoothly. It's actually quite astonishing how much we like repetition. We like simplicity and systems we understand," explains Reber.

Researchers have found that when a statement is repeated it will seem more true, and if a research subject has a cognitive leap and perceives the repetition even faster—fast thinking—it will be perceived as more true than before. This leap is one of the basic constituents of an "aha" moment. At the University of Basel, research subjects looked at a series of statements using colored words on more- or less-contrasting colored backgrounds. First they were shown low-contrast words a couple of times, which were difficult to read. Then the subjects were shown the same words with a high-contrast background. As the words became easier to perceive, a kind of "aha" moment occurred. The strange thing was that the statements they read were immediately perceived as being more true.

"The weird thing is that repeating a statement makes it seem more true, even if it isn't necessarily so. Similarly, an 'aha' moment will also seem fundamentally true," says Reber.

Some researchers wanted to explore this further and used false statements that included an anagram. For example, the research subjects were shown the following sentence: "There are more than 100,000 craters on the noom." Those who recognized the anagram—in this case, "moon"—were more inclined to believe that the whole statement was true, even when it clearly wasn't (there are just over 5,000 craters on the moon). This meant the researchers could prove that the *sense* of something being true increases when experienced in conjunction with a sudden insight. So you can easily believe something that is totally false, provided you have an "aha" moment at the same time.

In a similar study, people were shown a picture of an object (such as a camera) that was initially blurry but gradually became clearer. And as it became clearer, the researchers observed a positive effect via the activity in the smile muscle (the zygomaticus major)—a small Cheshire Cat–like smile—which coincided with the subjects suddenly recognizing the objects they were looking at. The researchers believe this demonstrates how the tantalizing sense of well-being you get during an "aha" moment is an automatic response.

"We like things that go smoothly, and 'aha' moments are moments of increased smoothness," says Reber.

Three things characterize an "aha" moment: it comes suddenly, it feels right, and it involves pleasure. You feel like the solution to a problem or idea has fallen into place or that what you were thinking suddenly makes sense. An "aha" moment feels effortless. You like it; finally, your story makes sense, or you see something in a new way.

"Aha" moments can be as varied as understanding an equation, or suddenly realizing that you want to be a vegetarian, or in the blink of an eye understanding that the Narnia books are based on the Bible. You are experiencing an "aha" moment when you realize that Thomas Anderson, the hero of the film *The Matrix*, has actually been trapped inside a computer-generated world controlled by evil aliens. Or you might understand a joke or a piece of wordplay. Think about how one audience member at a stand-up show might take a few extra seconds to understand a joke, how a light bulb will suddenly flash in their head, and the laughter will burst out revealingly late; it's like you can literally hear the cogs creaking in the back of their head before they get it. Jokes are in many cases "aha" moments.

"When I get an idea for a joke, it's quite a special feeling," says Laurie Kilmartin. Because a joke is the product of an "aha" moment, and Kilmartin—a stand-up comic and joke writer who has written for *The Late Late Show* and Conan O'Brien's show *Conan*—has a lot of "aha" moments. Her jokes also start with her feeling uncomfortable.

"I'll see a problem that no one has commented on yet," she says. "It'll be something unfinished and open that needs to be closed somehow. And this uncomfortable feeling can last; I'll go round thinking about it for a long time. But when you have the solution, it becomes protected and rounded, in a way; it's quite hard to explain," she says about joke writing.

It feels good. The discomfort is followed by a sense of relief. But a joke doesn't just fall into place because it's intrinsically *true*—even if it has all the characteristics of an "aha" moment, in that it seems to click, it produces a sense of relief, and it *feels* true.

"It's not the truth people find funny; nobody wants to hear that. I have to exaggerate things. And while there's nothing you can't make a joke about, you do have to consider the audience, because a joke is part of your interaction with them. What might be hilarious for one audience at a particular time might be totally wrong later. I had a joke that got huge laughs the first time I told it, and then never again, until I finally had to stop telling it instead of thinking that the audience was wrong," she says.

If the audience doesn't have an "aha" moment when they hear the joke, it's not a joke—there's no point to it. The "aha" moment of a joke will be very pleasurable; you'll laugh. It is not surprising that comedy is a huge industry, attracting millions of viewers.

But an "aha" moment can also be painful—like when you realize, for instance, that your partner's strange behavior is due to them having an affair. You suddenly see everything in a new light.

"There's perhaps not much pleasure to be gained from an 'aha' moment like that, when your partner has been cheating. Nevertheless, there's a sense of things slotting into place; they make sense, finally, compared to all the lies and inconsistent stories that didn't ring true. You end up with a balance between the pain caused by the dramatic insight of the infidelity and the comfort derived from understanding something that was unclear before," says Rolf Reber.

The same could apply if you came out of an exam and realized that you'd answered a question incorrectly. Obviously it's not pleasant realizing that you wrote a wrong answer, but your discomfort could well be overshadowed by the comfort of understanding the question in a way you hadn't previously.

We humans clearly like to understand, to create meaning, to make things complete, and to give context to stories that don't add up. This means that prejudices and conspiracy theories can feel very comforting too—if the facts aren't examined closely enough—because the comfort that comes from them being easy to understand provides a nice, strong feeling that they are true. So thinking fast can easily produce incorrect results too. An insight can feel so genuine and comfortable, so right and all-important, it can make you want to throw yourself from the Eiffel Tower with no fear of the consequences; you might be totally convinced you can fly!

The eight hundred "aha" moments that Rolf Reber and his research team collected have been categorized in several ways. By doing so, for example, they have seen that the men in the database experience "aha" moments mostly when they're on their own, while the women more often have "aha" moments when they're with company.

"Based on the material we have, it's also clear that many people have 'aha' moments when they are traveling, probably because they're seeing life from a new perspective. But that's something we do all the time; we look for patterns in everything around us. You look at the sky, and it resembles a face. You read the news and find conspiracy theories about the authorities. Looking for patterns is what scientific thinking is about too; religious and magical thinking is the same," says Reber.

Creating "aha" moments with other people is not as difficult as you might think; many Hollywood film scripts are full of "rewards" in the form of "aha" moments, made so that the audience will, ideally, feel good when they are understood. The nice feeling that comes from an "aha" moment is so powerful it can also motivate children to learn.

"There's a study from Canada that shows how pupils, who as a rule don't like math, become more positive toward the subject when they've had an 'aha' moment, compared with those who haven't," says Reber.

Now he is testing "aha" moments on children between three and eight years of age, and the studies will continue for the next few years.

"It's uncertain if children have any idea of what an 'aha' moment is, but they still want them, and we can test that by using videos and pictures. We don't really know what we're going to find and—for example—children have more 'aha' moments than adults. It's hard to say, because there are no descriptive studies that show how many 'aha' moments people normally have," Reber says.

Two weeks after speaking to the "aha" researcher, I am sitting on the edge of my four-year-old daughter's bed reading her bedtime story. Like most four-year-olds, she will use an unreserved amount of energy on acquiring sugar, in any shape or form. I am convinced she can smell candy from a hundred feet away. Her thoughts revolve around chocolate, lollipops, apple juice, and honey, despite her rarely getting any. On this particular night, she looked at me attentively while I sat, reading aloud from an illustrated copy of *Alice's Adventures in Wonderland*.

"Chocolate-milk-juice," she said proudly. "Wouldn't that be good, Mama?"

It was clearly an idea she found extremely comforting—she had finally managed to connect two things that tasted good, separately, into something totally new: something disgusting.

"I just came up with it now, in my head," she replied, self-satisfied and precocious, when I asked her where she got the idea.

As far as I was concerned, I was sure I had just witnessed an "aha" moment. Unfortunately, her knowledge of soft drink production was clearly so bad that she was unable to create a nice, new beverage with any market potential, but luckily she had no food industry contacts either, so she was unable to go into production.

I would love to have seen inside her brain just as she got her revolting idea. But it's, of course, impossible to connect directly to the brain of a person suddenly having an idea, since they normally come without warning and in quite unusual settings, far away from brain researchers. Nevertheless, researchers working with "aha" moments are keen to find out exactly what happens when someone gets an idea (be it a good one or not). So they have devised several tests, like those I described earlier, which create a type of "aha" moment when they are solved. For example: What do we associate with the following three words—cream, skate, water? (The answer: *ice.*) Or something trickier like being shown the incorrect sum 9 + 3 = 5 written in matchsticks, and having to correct it using one matchstick. (The answer: move the vertical matchstick in the plus symbol to the 5, making it into a 6. The sum would then be 9 - 3 = 6.)

9 + 3 = 5   9 - 3 = 6

Using functional magnetic resonance imaging (fMRI), researchers can scan the brains of research subjects while they are actively solving similar tasks. Because what happens in the brain when it is solving a task can be measured, fMRI allows the researchers to view, among other things, the flow of blood in the active parts of the brain. Those taking part in the experiment have to press a button the moment they find the answer, which allows the researchers to see what is happening the instant a person solves a task creatively and has a kind of "aha" moment.

A test at New York University showed that when the subjects experienced these artificially stimulated "aha" moments, the emotional center of the brain, the amygdala, becomes involved.

Even during trivial tasks, like identifying a hot-air balloon in a blurry photo, the amygdala will place the experience in the test subject's memory—via the brain's memory center, the hippocampus—the moment the test subject has an idea. An "aha" moment is emotional, and therefore memorable. In an experiment at the University of London, researchers were able to observe activity in the left temporal lobe (the area of the brain that solves complicated tasks) and also the thalamus (which sends signals from the sensory organs to the brain)—the dopamine system, the brain's reward system, was activated by a test subject solving a task they had been given.

But you don't really need a neurologist to explain the feeling of satisfaction that comes from having an idea, and it is far too simple to use dopamine as a reason for why we become addicted to things. Were I to tell you about my "aha" moments, you probably wouldn't feel anything yourself, because an "aha" moment that occurs spontaneously is intimately connected to an individual person and what they know and understand—you cannot swap your idea with someone else's good idea and get the same good feeling. A really good idea—that you are certain nobody has had before—feels like such an almighty kick that you might just want to run naked and cheering through the streets.

But what happens before the idea hits you? You're perhaps walking round with a problem on your mind, something you want to solve, something that's bothering you: a royal crown that might actually be made of silver, or skis that don't work properly, or a vague sense that you have to write a novel or understand something, even though no one has asked you to. Large and significant "aha" moments come totally without warning, and do not show up in laboratories or fMRI scanners.

One of the most important ideas of modern times was conceived after years of pondering a problem no one realized they had. A sixteen-year-old boy walking to school in Zurich one day began to wonder about something that would change the way we perceive time and space. He had a sudden idea: What if it were possible to travel at the speed of light—would you be able to see yourself in the mirror? Would light reach you if you were traveling at exactly the same speed? Ten years later he had not forgotten the problem, and while working at the patent office in Bern, he met his friend Michele Besso and happened to discuss it with him. Albert Einstein later wrote that this conversation triggered a storm within him. The solution felt like a huge relief: he suddenly realized that time is not absolute. The following day he met Besso again and, without saying hello, just said, "Thank you. I have completely solved the problem." It was the spring of 1905, and Albert Einstein was about to change everything we knew about the universe, time, and energy with his theory of relativity.

The American scientist Richard Feynman, who won the Nobel Prize in Physics in 1965 and was involved in the development of America's top-secret atomic weapons program during World War II, became motivated in his research primarily by how wonderful it felt to get a good idea.

"Absolute ecstasy. You just go absolutely wild," he would say later.

One of Feynman's important discoveries was related to a theory about liquid helium—and how the gas becomes a so-called superfluid. He had been working on the problem for two years, at which point he looked up from his papers and said to himself: "Wait a minute—it can't be quite that difficult. It must be very easy. I'll just stand back, and I'll treat it very lightly. I'll just tap it, boomp-boomp." And the solution came to him.

But Feynman was never able to re-create this wonderful feeling, and he dreamed about it for the rest of his life: "So how many times since then am I walking on the beach and I say, 'Now look, it can't be so complicated.'"

It turned out that *forcing* an "aha" moment on command wasn't that easy. Think about it: Have you ever managed to sit down and plan an "aha" moment? Have you ever managed to squeeze your eyes shut and just get an idea from sheer willpower? Normally when it strikes you, you'll be doing or thinking about something entirely different. You'll be standing in a forest and the Cheshire Cat will suddenly appear.

While the "aha" moments of scientific pioneers like Einstein often seem like great moments of insight, "aha" moments in art and literature are generally smaller in scale. Instead, these artistic "aha" moments can come in long strings of insight that will, for example, make an artist repaint part of a picture several times, claims the writer William B. Irvine in his book *Aha! The Moments of Insight That Shape Our World.* This is because a work of art often contains numerous small problems that need to be solved consecutively.

"To find this truer and more fundamental character, this is the third time I'm painting the same spot. Now that's the garden that's right in front of my house, after all. But this corner of a garden is a good example of what I was telling you, that to find the real character of things here, you have to look at them and paint them for a very long time," wrote Vincent van Gogh to his brother Theo, consumed with the problem of how he could render the world around him as effectively as possible in strong, clear colors. On this occasion, he was referring to a bench.

Similarly, a writer's understanding of their work will continually grow as a story evolves, each chapter filling with new ideas and each twist representing a new collection of ideas, reflections, and "aha" moments. But just referring to scientific ideas as *big,* and artistic ideas as *small,* isn't really enough. We need to perhaps differentiate between the huge, paradigm-shifting ideas—like the theory of relativity—that changed everything we previously knew, and small ideas that occur within a system, like the ideas you might get while painting a picture. Some occur within a specific area, like if you're working at a soft drink factory and get an idea for a new product, and some are a giant leap into the unknown, like when you get an idea for a great new novel.

"When I was a child, I dreamed about being both a writer and an actor. But as an adult, my dreams were a bit more levelheaded," says the author Maja Lunde, who I visited at her small, terraced house on the east side of Oslo. Lunde is about to move into a new house, only a few hundred feet away and a tiny bit bigger, which she has built using the money earned with her creativity. She likes the area for its light and space, its views of the city, and its close proximity to the forest.

"I was actually planning on becoming some kind of cultural bureaucrat. Then I applied for six months of unpaid leave, on top of my maternity leave, from the PR company I was working for. After that I began writing scripts for film and for Norway's children's TV channel, NRK Super. And the script writing just took over completely; it was like returning to something I had done as a child. I didn't go back to my old job, of course, and in the end, I just quit," says Lunde about how she became a full-time author.

"I started writing fiction out of frustration with the film industry. Everything took such a long time; the scripts never came to anything—at least, it took a very long time before they did. So I needed to find a way back to the feeling I had as a child, when I just lived in a world of my own. But I never thought I would earn a living from it. I'm not exactly a risk-taker, and to live off of writing isn't a particularly secure job," she says.

When she finally took the plunge, she had three small children and a mortgage, so she wasn't exactly yearning for a wild bohemian lifestyle. But she knew that she had some good ideas. She began by writing the children's book *Across the Border*, about the smuggling of refugees into Sweden during the Nazi occupation of Norway—and then she had an idea that would turn her life upside down. The novel *The History of Bees* has been a bestseller in both Britain and Germany. All because of an "aha" moment she had in a quiet Norwegian suburb. The idea, which changed her life, came to her after watching a documentary about climate change and the extinction of bees.

"I'd been wrestling with all these stories that were closely related to my own life, something we have quite a tradition for in Norway, but I just wasn't engaged by them. You have to write about things you're really passionate about, and the routine challenges of my life don't fit that criteria," she says.

When she switched off the TV, she knew she had the idea for her first novel, an idea that had emerged while she was learning about global warming, pollination, and bees—it was almost like it buzzed out of the TV and became something bigger—a huge story. She could see the whole thing, three stories from three different times and places that could be woven together into one novel: a story about a British beekeeper during the 1880s, an American beekeeper who witnesses the extinction of bees in 2007, and a story set in China in 2098—many years after the bees are gone. Lunde immediately abandoned her other ideas because this was a book that just had to be written. When *The History of Bees* finally came out in 2015, it won the Norwegian Booksellers' Prize and became an international success; at one point, it was selling to the value of 200,000 euros a week in Germany alone, and it stayed on the German bestseller list for months. Now the book has been translated into forty languages. Realizing that she needed to write more about the climate crisis, about the destinies it involves, and about the future of the planet, Lunde then got the idea for a full series, which she called the climate quartet. The third novel in the series, *The Last Wild Horses*, came out in 2022.

Lunde continually gets new book ideas, especially when she's not expecting them.

"I have a lot of projects that I've not yet completed, ideas that I'm saving and will return to and write one day. There's a children's book, and a novel about death. I've written outlines for them, but I need to find the right angle or approach. So they'll just sit there until I can figure out a way of doing them," says Lunde.

When she is out jogging, walking, or cycling, she will try to avoid listening to music or audiobooks and just allow herself to be enveloped in silence. Then, a picture will suddenly appear in her inner eye, and she'll know that it has to be examined more closely, to see where it leads. "Imagination is a gift. It makes us human, and it's something I'm extremely grateful for," she says.

Lunde's work also involves a great deal of "what if?" thinking, and the most important of these questions for her right now is: What if we enter a climate crisis because our goal of keeping global warming below two degrees has not been achieved?

Now she is writing the final book in her climate quartet. It is a complicated structure to weave together because, in the end, all four books must interlock to create an even bigger patchwork of events. All this requires an exceptional ability to see the overall picture, a structured brain, and a lot of "aha" moments. Lunde incorporates the writing techniques she used during her years as a scriptwriter; she has to create a mental map of the events, and works in an orderly and systematic manner, often on a plane traveling abroad to one of her book launches. But while her books fly off the shelves around the world, she, on the other hand, lives a modest life in east Oslo typical of any mother of three. She has cut out all unnecessary shopping, takes the family on holiday by train, and volunteers at the school flea market, which raises funds for her sons' marching band. Her life is far from being all parties and champagne and private jets. But her success means that she can write free of any immediate financial pressures, and knowing that she has an audience waiting for her next book.

"Were there many ideas buzzing around in your head when you were not a professional author?" I ask her.

"As a child and teenager, yes. I didn't have time when I was an adult. But I was restless; I didn't like the feeling of churning stuff out. Deep down, I knew that I didn't just want to leave a mountain of press releases behind when I died," she says.

Before becoming a writer, Lunde wrote a lot of press releases—and although it was a job that had to be done, she quickly realized that it wasn't quite enough. She likes to explore and to learn while she is writing, and devotes a lot of time to acquiring the relevant knowledge for a given project. Which brings us to an important point: one of the most fundamental elements of creativity is *curiosity*. Without curiosity, our world would never have grown; if we don't learn anything new, we won't get new ideas. Everything starts there. Questioning and probing and wondering has led humankind from the Stone Age to the Space Age—from the turn of the first wheel to the birth of the internet. Curiosity, combined with our ability to study and collaborate, to think up good ideas and construct great stories, has changed the world so much that we are actually in the process of destroying the foundations of life.

"There's something about people. We have this insatiable need to own everything, and curiosity is a part of that. Curiosity and creativity, our ability to see new connections, are crucial to science," says the biologist Markus Lindholm, research manager at the Norwegian Institute for Water Research (NIVA). Lindholm, who previously wrote a book about evolution, is now writing a book about curiosity.

Lindholm wants to show how science is driven by a fundamental sense of curiosity, without which we cannot progress.

"Science is all about fumbling in the dark, asking new questions and having crazy thoughts," he says.

One might easily see curiosity as being synonymous with wonderment, but Lindholm believes the two things are essentially different. One leads us toward silence and poetry, the other toward more concrete knowledge.

"To wonder means to stand before creation and become filled with something existential—while curiosity has a direction. And to get a direction, you need to have knowledge. You work your way into a tiny hole, using highly specific knowledge—particularly when you're working scientifically. After that, when you have passed through the hole, it opens up, and you can attain a sense of full-blown wonder. Childlike wonder doesn't get you that far, but that's where it begins. It fuels our curiosity," says Lindholm.

Research on curiosity has shown that to have some knowledge radically increases our need for more. If you know nothing about something, you are probably not very curious about it either. This is where *openness* comes in. Openness means asking and inquiring about things when you don't even need to, and wondering about things that can make you look naive. If you are open, then you dare to be inquisitive, even if it sometimes involves being wrong.

But the concept of curiosity has a more winding history. It's as old as the beginning of the Bible where Eve, out of curiosity, reached for the apple and immediately assured her fall from grace. Not such a good idea then, being curious, when all it took was a woman's curiosity to banish humankind from Paradise! In the fourth century, the philosopher and Christian theologian Augustine classed *curiositas* as a sin, saying that it turned us away from the pure contemplation of God. Curiosity is directed outward at the world around you, instrumentally, and leads invariably to temptation. "Curiosity killed the cat," we say, as a reminder not to poke our noses where they don't belong. In the late fifteenth and early sixteenth centuries something changed. Not that people weren't creative in the Middle Ages—consider the amazing cathedrals, the beautifully illuminated manuscripts, and the elaborate arguments for God's existence—but during the Renaissance what they idealized was the individual. People became more curious, and the world became a stage where humanity could have a leading role. And the Renaissance (meaning "rebirth")—so named because of how the Greek artists and philosophers were rediscovered during this time—was about more than simply reproducing the past. The carnival never ended, and the future was open. The people of the European Renaissance felt entitled to own the universe, to conquer and control it in a whole new way, from the remotest stars to the innermost cells of the body. In 1596, at the world's first university in Bologna, an anatomical theater was built where the students could observe dissections and see inside a human body, something that before then had been as impossible as crossing the sea to a faraway continent. At the same time, Copernicus's and Galileo's observations of the universe showed that the modern world was no longer magic. The earth and humankind were no longer the center around which the sun revolved—and eventually the Garden of Eden ceased to be a geographical point on the map.

Around the year 1600, the artist and rebel Caravaggio painted *The Incredulity of Saint Thomas,* in which all the Christian warnings against curiosity appear to have been turned on their heads. In the Bible, Incredulous Thomas (or Doubting Thomas) says he cannot believe that Jesus has risen from the dead before he sees it for himself. In Caravaggio's picture, Thomas has become the main character, the star, investigating the mystery with his bare hands. He has become the role model of the modern scientist or artist, for those who will not allow curiosity to rest and who want to find out everything for themselves, independently of the authorities or religious revelation.

It starts with curiosity and continues with knowledge—and knowledge leads to great ideas that change the world. Good ideas are, as we know, far easier to get if you already know a lot about something than if you don't.

Curiosity has simply accelerated in our culture. It has gone from being a sin to being an ideal.

Trude Lorentzen's job is based entirely on curiosity. It is also specifically connected to the modern world, where our curiosity means we constantly need to know what's happening around us. Lorentzen is an award-winning journalist who has worked many years for *Dagbladet Magasinet,* one of Norway's biggest news magazines. She has also been nominated for Norway's prestigious Brage Prize for her book about her mother's suicide, *My Mother:*

A Mystery, and teaches journalism students all over the country. Curiosity, good language, and good ideas are her livelihood.

"A good idea is an idea that makes the other journalists slightly envious, because there's something immediate about it. The moment you hear it, you know there's something there; you have captured something. You've put words to something that everyone has been wondering, more or less unconsciously," says Lorentzen.

"When you present that kind of idea, something very special happens to the atmosphere in the room. Because everyone there has probably had that idea already, but they were too afraid to formulate it completely. Like the time I saw a fashion show in Milan and thought, 'Is it possible for ordinary people to dress like that?' before thinking that I should really just try it. It's something perhaps everyone has thought of, but I made it into a project and reportage."

Trude Lorentzen has to be connected to something bigger than herself, to the culture and the invisible network of thoughts and ideas connecting us, to understand what people are passionate about—and to connect the thoughts and ideas we already know. A good idea will make you look at something you already know from a new angle, without completely abandoning what you know, like getting the idea for a phone that will show moving pictures. When you hear a good idea, you think: "Of course. Of course, that's a good idea!"

"I still get scared that nothing will come," she says.

Because ideas don't necessarily come when we want them to. You can't always brainstorm them, and you can't just close your eyes and force them out either. Without ideas, it is difficult to do a good job as a journalist. A high-octane workplace like *Dagbladet Magasinet* is propelled by its journalists; it is they who develop the top stories, and it's those top stories that are the most fun to write.

Therefore, every day Lorentzen takes a shower. Not solely because she wants to be clean, but because the shower has quite a special role in the process.

"That's absolutely right! I get ideas in the shower," she says.

In Lorentzen's bathroom, there are no plans or demands, just solitude and warm running water. It's a mental break room where she cannot check her cell phone or the internet, and where her two children or husband cannot interrupt her train of thought. In the bathroom, she can make free associations until something comes to her, and then she'll often experience what Archimedes did in the bath.

When she gets an idea, it's important to mold it into shape.

"Many of my ideas just float around in the back of my head. I spend a lot of time with my own thoughts trying to formulate a response to the world around me. I have a natural inner monologue, and I think it's important to take my inner questions seriously. My philosophy on life is that the world can always be different—I have lots of 'what if?' thoughts," she says.

For Lorentzen to call something a good idea, two criteria have to be met: The idea has to sound catchy and immediately recognizable to other people. It should also be a little bit tricky to follow through—uncomfortable perhaps—or difficult, or just embarrassing, like going out to buy milk and bread at the local shop dressed in *haute couture*. If you can tick both those boxes, then you probably have a good idea.

One such idea was an article Lorentzen wrote about all of her ex-boyfriends, and how different her life would have been if she'd ended up with one of them, instead of the guy she now lives with and has children with.

"I actually wanted someone else to write the article, but none of the editorial team would do it. So I just had to write it myself. Everyone has their own version of that story, so when I pursued the idea, I wanted to show the deeply existential aspect of what everyone already understands: how life, and you as a person, would have been different if you had made other choices. It's something many people have thought about, but no one has spent three weeks of work figuring it out. This is where being difficult comes in—had it been easy, everyone would have done it, but in this case, I was the first to write it. And that gives what I write a little edge, because something is at stake. I had to be brave to make it happen."

Another idea Lorentzen had, which had similar universal appeal, was to write about all the empty children's bedrooms being kept as they were by parents all over Norway, just as her father does with hers. At first, it began as a pretty tame idea about celebrities returning home for Christmas. Not quite what she was looking for. But while she discussed the idea in an editorial meeting, something struck her: there were sixty-nine bedrooms just like this in Norway—all empty because of the 2011 terror attacks.

"Suddenly the idea acquired a totally new depth. I wanted to show the silence in those bedrooms and talk about who these young people had been when they were alive. It was a difficult article to write, and many of the parents refused. But many of them said it was a good idea, even those who couldn't bear to speak in person. So I couldn't give up."

In 2012, Lorentzen, along with her colleague Eivind Sæther and photographer Adrian Øhrn Johansen, won the most prestigious award a journalist can receive in Norway for this magazine story.

"There was no celebrating, because the whole story was totally heartbreaking," she says about receiving the award. "But the reportage had at least succeeded in showing who those young people were, according to those who loved them." Among the grounds for their decision, the jury stated: "The power of this story lies in the simplicity of the idea."

Clear and simple ideas are the driving force behind Scottish artist Katie Paterson's work. As an art student she had little more to show than a book full of ideas, since what she had been working with for years—all of her large-scale ideas—was just too difficult to implement. Now, many of Paterson's grand projects have become a reality. One of her pieces consisted of wood from every tree species in the world; another used images of all the documented solar eclipses in human history, printed on a large, dark disco ball. She also created a piece that let participants make a phone call to a glacier and listen to it melting. Her latest project, situated outside Apple's headquarters in California, is a series of glass columns made of melted sand from all fifty-eight of the world's deserts, from all the world's climate zones.

Another of Paterson's installations, *Earth-Moon-Earth*, transmitted the notes of Beethoven's Moonlight Sonata, via Morse code, to a receiver on the moon, which then "reflected" the notes back to earth. That lunar version was then performed by a self-playing piano, complete with periods of silence caused by the moon's craters and patches of space where there wasn't any signal. Paterson's art, which involves her close collaboration with scientists and academics, raises issues concerning time, space, and humanity's place in the world. And just as research on other people's "aha" moments has shown, Paterson's ideas are very memorable to her.

"I can remember getting every idea I've had. The ideas themselves will evolve, but the first seed is always very clear," she says.

"The Moonlight Sonata idea came to me in a cupboard in Reykjavík where I was reading up about the moon and found 'earth-moon-earth communication,' which involved a big old clunky computer from the early days of the internet. I was working in a hotel at the time, and remember taking a walk later that evening, past a graveyard, looking up at the moon, and having the idea of transmitting music using this old technology. The idea of having it played on the self-playing piano came later, but the idea about transmitting it came to me in Iceland.

"I got the phoning-the-glacier idea when I was a student. It's less clear exactly where I was, but I know I was reading about listening to dolphins using an underwater microphone, and that I wanted to do the same with a glacier! I remember feeling elated and telling some friends about it in class, and then writing to Ken who'd made the dolphin microphone."

One of Paterson's ideas is growing close to where I am writing now. A short streetcar ride away, in the forest just outside Oslo, are one thousand newly planted trees, which in the twenty-second century will be cut down and used as the paper in one hundred books. The idea, known as the Future Library project, involves a different author submitting a text each year, which will all be printed as an anthology in 2114.

Until then, the manuscripts are being stored in Oslo's new public library, in a room designed by Paterson with Atelier Oslo and Lundhagem. The room's curved wooden walls are interspersed with one hundred narrow shelves, one for each of the manuscripts, which will remain unread until they are published.

"I got the idea for the Future Library on a train to Whitstable, where I was working on a project. I was drawing tree rings there, and visualized the whole thing, but at the time the idea only involved one author writing a book that would be printed using paper from the trees one hundred years later. The idea of using one hundred authors came after that," she explains.

Paterson has a disciplined work ethic and no time for chasing the mythical bohemian life of an artist; she has a little boy and big projects to supervise. As an artist, her most important daily routine is setting time aside for doing nothing—she creates space in her calendar for it—to allow new ideas to form. Unlike many other artists, who might become inspired while painting or working with clay or stone, Paterson's most important raw materials are simply good ideas.

"What's fantastic about working with art is that I can work with anything—and that I often work closely with scientists. At school I was really good at math and I enjoyed it, but I have always loved drawing, so there was never any doubt that I would become an artist. I learn something from every single project, whether it's new information or a new skill—perhaps a skill I might never even need again, like casting scented candles, or transmitting signals to the moon, or making glass out of sand."

Some ideas are more difficult to actualize than expected, and require huge amounts of preliminary work and research. Paterson recently made a "ticker clock" that records and displays the total number of sunrises there have ever been—a pretty megalomaniac idea.

"The existing number was even more difficult to calculate than we'd imagined," she says. "It wasn't just something you could find using Google, partly because we don't know how fast the earth rotated in its early years. So the scientists spent two years factoring this into their calculations before we had the answer."

Getting a good idea can often feel like luck, but luck and curiosity go hand in hand. And it was pure luck and chance that gave us one of the most important discoveries of modern medicine—one that might never have been made had Alexander Fleming been a more hygienic scientist. One summer day in 1928, some fungal spores flew in through Fleming's laboratory window and landed in a petri dish, where they remained for the rest of the summer. It wasn't until Fleming returned from his vacation that he saw what had happened; during the summer, the tiny spores had killed the bacterial cultures that had been living in the petri dish. In 1945, Fleming was awarded the Nobel Prize for this discovery of one of the world's most important medicines—penicillin. Without coincidence and luck, this discovery would not have been made, and without Fleming's openness and curiosity, we wouldn't have benefited from it either.

Kaja Gjedebo, a renowned jewelry designer who is now breaking through internationally after launching a collection in Germany, says that one of her greatest creative achievements came as a result of luck and untidiness. To be more precise, her own untidy desk.

"Nothing I do ever feels like it's been wasted. It always leads to something, although I don't always know what it is. For me, creating something is about making myself available to chance. It's like kissing a frog; you might get a prince out of it in the end. That's how it started—I had put some parts on my desk, and there they were," she says.

Gjedebo lives in a so-called "artist's home"—a 1950s-built dual-purpose house and workshop, with large windows that fill the place with light. But instead of the clean, minimalist surfaces you might expect in a functionalist building, every workbench is covered in all sorts of mess—fragments of metal lie everywhere, along with a random chaos of paper, tools, and other bits and pieces.

"Sometimes a botched project is the best thing that can happen! I was actually making a ring once, drilling and fixing, but it didn't work out, and I just left the pieces lying there. Then one day, I tried using those pieces on myself, as earrings. Now they're my biggest-selling design. By sheer coincidence, they fit perfectly around your earlobes and they lift them somehow—it's a particularly good design if you have quite long holes in your ears," she says while handing me one of them. You would think it had been designed as an earring, not just part of a ring.

A similar thing happened with a pair of earrings that didn't turn out how Gjedebo wanted: she welded the two parts together to form a ring and the piece ended up being extremely popular with her customers.

"It can take years for an idea to mature and become a piece of jewelry. I have so many different projects going on simultaneously. Whenever I tidy the workshop, I'm always excited about what might happen. I'll find something, look at it, and think, 'No, not today.' And suddenly, I'll see the road ahead," she says.

Gjedebo's "aha" moments are entirely physical; they happen when objects are just lying around—pieces will suddenly lie at an angle she has never seen them in before, or be placed strangely on top of another. For her, untidiness is a prerequisite for having an "aha" moment. There's even research showing that untidiness is good for creativity. In 2013, researcher Kathleen Vohs of the University of Minnesota's Carlson School of Management concluded that an untidy desk was good for all types of creative work. The finding caused a media frenzy, and disorderly creative types the world over breathed a sigh of relief; now they didn't have to tidy up at all! However, a more recent study from 2019 by a group of researchers in New York showed that untidiness has no significant effect after all, which shot a hole into the first result—showing, above all, that we have to be cautious when it comes to new research, and perhaps with untidiness too. But for Kaja Gjedebo, a messy workplace is a godsend and essential to having good ideas—research or no research. Her jewelry is as simple and elegant as her desk is chaotic.

"What the research on creative people shows is that they are not afraid to fail. They are humble. They have this huge inner drive to learn and explore, and they are very curious," says Joy Bhattacharya, professor of psychology at Goldsmiths, University of London—a university well known for its creativity, with alumni that include the artist Damien Hirst. Bhattacharya's research is based on the work of Joy Paul Guilford, one of the founders of creativity research.

Guilford, who was an expert on IQ testing, had tested the IQs of both children and military personnel, beginning in the late 1920s and throughout World War II. But something didn't add up; there was something missing from the tests. Then he had an idea: he would make a test of his own. In 1950, as president of the American Psychological Association, he delivered a speech at the opening of its prestigious annual conference. During the speech, he spoke enthusiastically about creativity, and triggered a flood of interest in the subject. Guilford believed that standard IQ tests were unable to identify people who thought differently than the normal population because they only really measured a person's ability to think conventionally, following a logical correctness—not unconventionally, irrationally, and creatively. His concept of "divergent thinking" involved thinking differently and seeing many possibilities, while an IQ test simply looked for one answer to each question: the correct one. He believed that many of the most creative people could still score badly on standard IQ tests.

This was a turning point in creativity research. Neuroscience was still in its infancy in the 1950s and had only begun scraping the surface of how the brain functions. One of the most significant figures in the history of neuroscience is the now well-known patient Henry Molaison (previously known among researchers as H.M.), who, in 1953, had both of his hippocampi removed during an operation related to his epilepsy. The results of the operation helped researchers understand many of the mysteries of the brain, particularly memory. With his hippocampi removed, Molaison was unable to learn anything new, which immediately demonstrated something quite crucial about how the brain functions. This meant that researchers could start mapping the function of the hippocampus, based on everything Molaison was unable to do and, despite the damage, everything he *could* do. Until he died in 2008, Henry Molaison cooperated with researchers through a range of tests and experiments, allowing them to study how his brain worked after the operation. What they found was that long-term memory is very different from short-term memory (or working memory), because Molaison was able to recount his life up to a point three years prior to his operation, which showed that memories have to be consolidated—massaged into our memory—over a long period. He could hold a conversation until something distracted him, at which point he would instantly forget everything he'd been talking about, because that is how our working memory functions in extreme situations. He was able to learn motor tasks, although he could never recall how, which proved that we have a separate motor memory—but he was unable to orient himself and had no sense of location. It is possible that he was also unable to remember anything vividly, sensually, or emotionally (characteristics of so-called episodic memory), because he would recount his memories in quite a flat manner. And after the loss of his hippocampi, he was unable to attach new episodic memories to his brain. From all these building blocks, memory researchers were able to compile a huge catalog of knowledge about this one specific, creative attribute of the brain. You could say the hippocampus is kind of a memory coordinator, retrieving and orchestrating memories—but most of the time it functions like the world's worst amateur theater director, and never follows the script. Our memory is a narrator, not a video recorder, largely controlled by our emotions and associations.

What memory researchers have learned since Henry Molaison's operation is in many ways disheartening for those of us who thought we could rely on our memories. Memory compresses and expands what we think we remember, and this changes still further when we experience new events in our lives. Traumas adhere more easily to our memory than fleeting moments of everyday happiness do. We recall our wedding day quite differently after getting divorced, and if we witness a crime we are unlikely to recall much of it correctly. Our memories are perpetually new stories about ourselves and what we know about the world—and are highly unreliable. We are all inherently creative.

But although memory research shows that our memories are governed by creative forces, it is not here that *creativity* research has its roots. It began of course with tests. In addition to IQ tests and Guilford's new creativity test, we now have personality tests—and the most popular of these among the world's psychologists and neuroscientists has been the so-called "Big Five," which measures five major personality traits: openness, conscientiousness, extroversion, agreeableness, and neuroticism. In addition, there are a total of thirty different measurable subtraits within the five major ones. Researchers assume that we all have some of these major traits, to a greater or lesser degree. It is also possible for someone to have all five key personality traits.

Openness is perhaps the most important trait for people who are creative. It implies that people who earn a living from being creative are, by nature, curious and engaged—and can provide long and unexpected answers to questions like "What can a brick be used for?" or "What would happen if gravity were suddenly half as strong?" Guilford developed a comprehensive test that would measure specifically this type of divergent thinking (i.e., thinking that leads in many directions) called the Alternative Uses Task (AUT). Later this was developed by the researcher Ellis Paul Torrance and used in the Torrance Test. In addition, there is Sarnoff Mednick's Remote Associates Test (RAT) for mapping convergent thinking (thoughts that are related), which measures strange associations to words.

"Of course, to define creativity using these tests is problematic, but it is the most widely used method for enabling us to quantify creativity and in turn make it into a field of research. This way, we have been able to build the research field brick by brick," Bhattacharya says.

Professor Bhattacharya's research shows that most people can become more creative using creative training programs (i.e., practicing various types of divergent thinking). But people who are particularly creative already will find these exercises less useful.

"This is most likely because people working creatively, at a high level, already have their own creative system, which a new training program is likely to disturb. If these individuals are experiencing stress and depression, we know that their creativity level will most likely go down. So they are vulnerable," he says.

After Guilford opened the door to creativity for researchers within psychology and neuroscience, a rush of practical applications, creativity exercises, and techniques emerged. You can just imagine how in demand creativity was just a few years after World War II, when much of the world lay in ruins and the survivors had to rebuild their surroundings and themselves. Creativity was essential to creating a new and better society.

But the wind had blown in this direction before Guilford. In 1942, the advertising executive Alex F. Osborn published the book *How to "Think Up"* in which he coined a new and groundbreaking technique that you will certainly have heard of: "brainstorming." *Applied Imagination,* which Osborn published in 1953 and which became an immediate bestseller, included tasks for stimulating creativity, such as: "Suppose you were a manufacturer over-stocked with tooth brushes. For what uses (other than brushing teeth) might you try to market your surplus inventory?" and "What new and additional functions can you think of for a helicopter?" and "To what uses might the shells of cocoa beans be put?" Osborn traveled around, teaching in schools and the military, and in 1954 he founded an institute that taught his "Creative Problem-Solving Process." During the hippie era of the 1960s and 1970s, the celebration of human creativity reached new heights, and both then and since, countless books have been written full of exercises aimed at making readers more creative. On Amazon, I found seventy thousand books on the subject. "Creative" is, after all, something everyone wants to be! But many of these books on creativity are really quite paradoxical, because following a creative method to the very last detail quickly defeats its purpose. Rules and norms can impede creativity. It also turns out that Osborn's "brainstorming" is not so conducive to creativity either, because a person can easily wind up following their group and end up with slightly conformed ideas. We get weird and exciting ideas far more easily when we are alone, according to new research on the method. It is no coincidence that Einstein didn't "brainstorm" the theory of relativity.

There is also another problem: Guilford's test was intended as a criticism of the definitive answers provided by IQ testing. What he ended up with in return was a new group of "test winners." But can a test really distinguish between what's creative and what's noncreative, and can we really categorize people based on such a test? Is it possible to come up with a blueprint for what's creative? Given what we know about memory's creative and associative nature, aren't all humans quite creative? And if we're all actually creative anyway, how can we unleash all this creativity?

What we now know is that the White Rabbit can pop up when you least expect, to lure you into an absurd imaginary world. Try and outwit the Cheshire Cat, and he'll vanish with a large grin. *Alice's Adventures in Wonderland* is one of my favorite books because it is so strange, funny, and creative in every respect, and at the same time so nonmoralizing—considering it was written at a time when children's literature was supposed to be morally edifying above all else. And this entire crazy universe popped up in the mind of Lewis Carroll while he was rowing a boat through Oxford, along with the children of the dean of Christ Church, where Carroll had a fairly *un*creative job as a math teacher and church deacon. On this quiet July day in 1862—when the last thing he had on his mind was writing one of the world's greatest bestsellers and something that would become a cultural icon all over the world—Lewis Carroll had an "aha" moment. He created a modern classic on this single rowing trip with the Liddell children, one of whom was called Alice; when the children nagged him to entertain them with a story, he told them all about what happened when Alice, along with her big sister, sat by a riverbank feeling bored one summer day. You can safely say that the Cheshire Cat appeared that day, with a big smile on his face.

When it was published in 1865, the first two thousand copies were misprints, so Carroll decided to give them away to hospitals and various homes. In his diary, he wrote that he hoped to sell two thousand copies of the reprint so that he might break even.

"Any other further sale would be a gain. But that I can hardly hope for," he noted. One of the discarded misprinted books was offered for sale by Christie's in 2016, with a price tag between two and three million dollars.

When the second edition came out, the first five thousand copies were torn off the shelves, and the book has been in print ever since. It was one of Queen Victoria and her children's favorites, and was read by a young Oscar Wilde. Since then, it has been translated into 174 languages and sold millions of copies.

In 1937, a television version of *Alice's Adventures in Wonderland* was made for the very first time. It was the same year that television started becoming more widely used in England—and it was also the year that everything turned upside down for John Logie Baird.

In 1928, Baird became the first person to broadcast television images in color and, in the same year, the first to make a transatlantic broadcast from New York to London. But in 1937, he was slighted by the company owned by the Italian physicist, inventor, and Nobel laureate Guglielmo Marconi—the same company that in 1925 had been so indifferent to his invention and wanted nothing to do with him. Marconi-EMI eventually managed to build a television that was less mechanical and cumbersome than Baird's, and which was better and easier for the BBC to operate. Baird had gambled everything on his brilliant idea, and had been providing television solutions to the British broadcaster for nine years, but now he had to consider himself beaten and outperformed. Nevertheless, he continued to improve and develop televisions, and pioneered both HD and 3D-TV technology. He also invented a video system, an infrared light, and a kind of radar. His contribution to television history has, in posterity, been justifiably praised.

Now, more than one hundred years since Baird had his peculiar idea in Helensburgh, I can watch one of his movies—the flickery black-and-white broadcast of *The Man With the Flower in His Mouth* from 1930—on the screen of my smartphone. Strangely enough, John Logie Baird's idea of *seeing through the telephone* has become more of a reality than he could ever have imagined.

2 | The Mad Hatter's Tea Party

OR: I LEARN TO BE SPONTANEOUS AND FOLLOW MY INTUITION.

"It's the stupidest tea-party I ever was at in all my life!"

ASK ANY ARTIST, musician, writer, or researcher what holds them back and you will realize that there is an invisible third person in the room. Some refer to this third person as "the judge" or "the saboteur," but as a rule it is known as "the inner critic." The best-known depiction of this negative force is perhaps from a cartoon: Donald Duck, when an angel and a devil appear on either side of his head, offering both good and bad advice on what he should do. The duck's creators probably stole the idea from Christopher Marlowe's theater classic *Doctor Faustus*, written in 1592, where two figures—an angel sent from God, and a demon sent from the Devil himself—fight over Faustus's soul. The winner in Marlowe's piece is the Devil, the inner critic, and Faustus winds up in Hell. In modern times—and perhaps thanks to Donald Duck—this image of Christian morality has been transformed into a depiction of the forces we all struggle with, between the evil intentions and high-minded ideals we sometimes find wrestling inside us.

The inner critic gives you negative feedback on what you're doing, and can be quite an unpleasant voice to fight. You can be sure that for every word hammered into my keyboard, I have had a comprehensive argument with my inner critic. Every step I take through this book is like wading through a quagmire in rubber boots; it is hard work, it goes "schloop," and I have to drag my feet. But I keep going regardless, mainly because I'm extremely stubborn and, for some incomprehensible reason, refuse to give up.

You may think that all my talk of inner critics is just attention seeking, but it's actually quite logical, not something we—those of us whose livelihoods depend on being creative—just make up. We really do have one, all of us. The inner critic is actually a pretty smart invention. It stops you from making an idiot of yourself. It shouts, "Off with her head!" whenever you come up with a terrible idea, and actually just tries to look after you. I named my inner critic the Queen of Hearts, after Wonderland's tyrant in chief (who bears more than a slight resemblance to the short, squat, and all-powerful Queen Victoria, who reigned over the British Empire when Lewis Carroll wrote the book). She is powerful and slightly mean, perhaps, and a representative for the rules and norms of society. However, being creative—like Alice—you need to break the rules occasionally in order to create something genuinely interesting. Because you are curious! And imaginative! And have strange "aha" moments! And sometimes you want to make something new and avoid being conventional and boring. If this sounds like you, then it's most likely that the Queen of Hearts comes running after you shouting, "Off with her head!" But you have to be like Alice: drink from the bottle that makes you big, grow all the way up to the roof, and pick up the angry little queen by her dress and dangle her in front of your nose.

I think about my daughter, and her chocolate-milk-juice. No inner critic there. A real Einstein, except with none of Einstein's knowledge or experience. Immediately after crashing into the bridge, I too felt briefly like a four-year-old, equally uninhibited. But little by little, my inhibitions returned, and this time with a vengeance. It was like waking up after a yearlong party and thinking, "Was I dancing on the table in my underwear singing hits from the 1980s? At lunchtime, at work?" (I didn't do that, of course, but that's how it felt.)

One day the queen banged on my door. "Hello!" she bellowed, "I am the Queen of Hearts! Your desk is a mess, and research shows that a messy desk is *bad* for creativity! Look at those ugly shoes! Have you looked in the mirror lately? And what gave you the impression that *you* can write? Nothing you write has any significance!" I nodded, dumbfounded. This meant war. If I was going to write a single word of this book, I would have to deal with her.

So where does she come from?

When we are born, our brains are ready to accept every single impression we experience. And our 100 billion brain cells—neurons—quickly form an incredible number of connections: synapses, as they are called. We can safely say that this is an extreme form of openness. But not everything can continue to be possible; at some point your brain has to start providing ready-made answers to a range of questions. This is crucial for learning. You will know automatically that the transparent container in your hand is a glass, and the white liquid sloshing around in it is milk, without too much thought. We learn words for the things around us, and for the behavioral conventions within our own culture—a process that begins when we're about two years old and continues until we are adults. The technical name for this is "pruning," where the connections become fewer, but stronger—a network of tiny pathways within the brain that become roads, then avenues, and finally eight-lane highways. We become socialized; we learn what is important and unimportant.

We talk about highways in the brain because the synapses become reinforced in a particular direction. This transformation from signals to superhighways is called long-term potentiation, and was discovered by the Norwegian scientist Terje Lømo in 1966. We don't want to simply repeat what we already know; we constantly learn new things as we grow up. The brain consolidates our memories, cutting and pruning all that we don't need. This creates a memory network you can build on when learning new things as an adult. When you follow a new and unexpected association or idea, it doesn't involve choosing the shortest route, from one point to the next; it means going from A to B via Q and X. Maybe you won't make it to B at all, once you are out on this new road? We cannot always move around via strange associations, of course; we also have to follow conventions and habits, to stay connected to each other and our culture. And this is where *she* enters the picture: she, the Queen of Hearts, represents socialization within society. For example, it was she who made us take it for granted in the 1400s that "the sun revolves around the earth," and, today, that "the earth revolves around the sun." It is she who reminds us of codes of politeness, such as saying "Thanks for having me," or looking presentable when you meet someone for the first time. The inner critic makes sure we have control over what's good and bad, what's acceptable and unacceptable, what's nice and what isn't so nice, and how the world—and culture in general—is held together.

"We can determine the point when a child starts to develop an inner critic. It happens roughly after the age of six, when children have internalized a personal standard for what a good drawing is, and evaluate it regardless of what their parents say about it," says Evalill Bølstad Karevold, a developmental psychologist at the University of Oslo.

Before six, children will show pictures they have drawn to their parents without thinking about the response they might get, but eventually they will become more concerned with the feedback. And in the end, it might not matter what the parents think, since the child will be relating to their own inner standard. The drawings will still be "ugly," no matter how much the parents insist that they are beautiful.

"Obviously, it can be good for a child to internalize criticism like this, but it can also shut the door to creativity. The main focus of my work is the temperament of the child. Every child is different, and some have a far greater fear response than others. Some have more rigid mannerisms than those who are more outgoing, open, and flexible. How the parents respond to the child in this situation will affect how the child responds later in life. It involves both genetic and environmental factors," says Karevold.

Being a parent means identifying this temperament and working either with or against it throughout the child's upbringing. If your child is outgoing and funny, you can provide them with something else, like structure and routine. If your child is introverted and more safety-seeking, you can encourage them through activities that explore their spontaneous and outgoing sides.

"This will lead to the child being more flexible, and better equipped for dealing with life's tasks and challenges. And being flexible is a part of creativity. But the most important thing of all is to give the child a feeling of safety. This is the main foundation for being creative: a child should feel safe enough to explore the world, to risk making mistakes," she says.

Exposure therapy can make an anxious child or adult less afraid. It involves gradual exposure to something you find dangerous, in small doses, until it no longer feels scary. To work creatively actually means you are constantly testing those boundaries. According to researchers like Joy Bhattacharya, a creative person is someone who has maintained some of the strange connections between the different areas of their brain, and who is not entirely conventional and predictable according to societal norms. Those defined as creative are unafraid of trying out things that are contrary to the generally accepted norms. It is likely that they have a bigger and more finely woven network of connections between the nerve cells in the brain, and a bigger than normal arsenal of associations—something that makes their brains more flexible, more robust, and, at the same time, more unpredictable. A bit like a child, perhaps.

"All children are artists. But very few of them continue to be. We streamline ourselves, and suppress what those of us in creativity research call 'divergent thinking.' We don't allow ourselves to think unusual thoughts; it's not something that is rewarded socially," says Bhattacharya. "To work creatively is to allow yourself to make mistakes. It involves having some kind of grit. Hard work over time will be rewarded. The idea of needing talent is overrated."

An experiment that involved music teachers and their students showed that the music teachers were unable to guess which of the students would become professional musicians as adults; the students who had a clear musical talent, but never practiced, never became especially good. Success comes through good interplay between talent and hard work—and between cultural codes and original ideas.

I think I might just try to get a more even balance between being totally free and being restricted. What works best seems to be a combination of stubbornness, hard work, knowledge of conventions, and original ideas. The philosopher Aristotle was perhaps right—Plato's student talked about "the golden mean" in all aspects of life. If what you are doing is to be of any cultural significance, you should retain some of your inhibitions; if you want your work to resonate with other people, you cannot be detached from them. It doesn't help to be a "mad genius." What actually happens when you're being creative is quite different: there is an exchange. The writer Suzanne Moore has revealed that she writes in bed, alone, wearing lipstick. "It is a sign I am moving between the inner and the outer world. That is what writing is, an oscillation between the internal and the external," she writes.

Personally, I spend far too much time arguing with my inner critic. So, I think I'll talk to a psychologist; perhaps he can get the Queen of Hearts to calm down a little.

"It's quite possibly just Freud's 'superego,'" says psychologist and writer Peder Kjøs, when I talk to him about the problem.

Freud divided the human psyche into three parts: the id, full of repressed desire and longings; the superego, which represents societal norms (which, during the early 1900s, were pretty strong in Vienna, where Freud worked); and the ego, which constantly mediates between the two, between our instincts and societal conventions.

"You need to ask yourself where this merciless inner critic is coming from, and what purpose it actually serves you," Kjøs says.

It serves a lot of purposes: when I listen to my inner critic I'm able to behave more or less normally, and it's more of a comfort than not; I actually feel quite weird in general. When I overhear the Queen of Hearts, all kinds of strange things can happen; I have far too many weird thoughts and jokes on the tip of my tongue at any given time—I can never say everything I'm thinking out loud!

"What's creative and what's free contrasts with what's disciplined and effective. An inner critic requires having a form of discipline. Order and belief in the future have brought us a lot of good things. In our time, however, we've placed a lot of importance on what's unique and individual, and artists like Munch have glorified madness and the forces of chaos. But discipline isn't necessarily an obstacle to creativity; the two just need to be balanced," says Kjøs.

Kjøs writes constantly, and is highly disciplined. In 2019 alone, he published three books and made a podcast and a TV program, in addition to contributing regularly to Norway's biggest newspapers. Yet he is seriously troubled by his own inner critic.

"Overcoming that voice isn't easy. I try to completely disconnect from it when I write, and then reconnect to it afterwards when I'm reading through the text. What I want above all is to find some kind of balance between order and chaos. There's a profound wisdom in Asian thinking. Kali is the Hindu god of both creation and destruction. They are parts of the same individual, not opposites," he says.

Kjøs believes, for example, that to accept that everything has already been said can make it easier, not harder, to write.

"It isn't possible to tell an original story today. What's different is the perception of something, *your* perception. It is about the eye that sees, and precisely *how* you see the world. So you have to focus on that," says Kjøs.

But for a while, a very long while, I had no perception.

After colliding with the bridge headfirst, my brain felt like it was continually bursting with energy and ideas. I was also quite rude and insensitive in the immediate aftermath. The concussion blunted my social intelligence (I've been told I have some normally) and my sense of humor; I was no longer able to tell jokes and make others laugh. I lost my rhythm, I lost my point of perception, I lost touch with other people. I lost myself a little.

I cried almost constantly after the crash. Not because I was sad, but because my brain was simply unable to distinguish between the impressions it was receiving. It was chaos, and not the kind of chaos that leads to great art. It was just... a mess. Most of the time I felt like a certain orange-colored former head of state: slightly out of control, slightly uncomfortable to be around, humorless, self-pitying, childish, and with a lack of self-deprecation—behavior that may have been due to me striking the right side of my forehead, near a part of the brain associated with inhibitions and sociality. Namely, the right temporal lobe.

In 2018, scientists in London moved a little closer to understanding the inner critic. They sent weak electric currents through the temporal lobe of a test subject who, at the same time, solved problems set by Sarnoff Mednick's Remote Associates Test. Mednick's test, along with the Torrance Test and Joy Paul Guilford's test, is the most recognized standard for measuring creativity. When the electric current was applied to the subject's head, they solved the problems in a far more original and less obvious or conventional manner than when they tried the same tasks without the electric current. All the test subjects were equipped with an EEG, and the current was applied so as to reduce the waves in the temporal lobe to alpha—a state similar to what you might experience during meditation, which occurs somewhere between dreaming and being fully conscious. This new research therefore attempts to replicate ancient meditation techniques!

During the interwar period, the German scientist Hans Berger discovered that there is constant electrical activity in the brain, measuring brain waves in cycles per second, or hertz. While resting with our eyes closed or during meditation—or in any type of conscious relaxing—our brains are in alpha waves, which are between 8 and 12 hertz. Delta waves (less than 4 hertz) occur during deep sleep, and theta waves (4–8 hertz) during light sleep, while beta waves (13–14 hertz) occur during conscious activity—used for concentration and focus and executive function. Gamma waves (up to 140 hertz) are the most intense of all. In 2014, Professor Mark Beeman, head of the psychology department at Northwestern University, found that when people solve problems and have "aha" moments, their brains are in the low-frequency alpha waves for a split second before switching to the high-frequency gamma waves—just as the test volunteers solved the problems they had been given. It happened every time! First alpha, just before, and then gamma as they solved the problem. Gamma waves tell us something about how new connections are created in the brain, while alpha waves are generally associated with divergent thinking and creative solutions—it was as though the subjects turned inward and made themselves relax just before having the "aha" moment.

At Queen Mary University of London, neuropsychologist Caroline Di Bernardi Luft has attempted to demonstrate how the creative process can be controlled and the inner critic restrained by stimulating the temporal lobe—an area closely connected to the frontal lobe, located near the right temple, just above the ear. To do this, she used alpha waves. Bernardi Luft is a senior lecturer in psychology and creativity scientist, and her team suspected that by opening up the brain to all sorts of associative leaps they could unleash a person's creativity, since the temporal lobe normally tries to block visual noise and extraneous sounds. Our working memory works hard to stifle any extraneous impressions so that we can concentrate on one thing at a time. But it also means inhibiting associations that are in the temporal lobe, so that we can remain more focused.

It was this inhibition that wasn't working terribly well for me after my crash. Which is quite possibly why it was so difficult for me to walk through an airport—I was simply unable to keep the sounds and visual impressions at bay. But we also repress mental leaps in the temporal lobe, and, although they disrupt our concentration, strange and unexpected associations are important for creativity—and after my accident, wild leaps of thought were things I had in abundance.

You could say that when the executive network is functioning, the temporal lobe works almost like a pair of blinders, making you more focused—much like a horse pulling a cart through a chaotic marketplace; however, it also blocks a great deal from your field of vision. With alpha waves, your brain adjusts the blinders to increase your field of vision—along with all the pros and cons that that entails—so that you no longer see the obvious path, and your brain has to find other, new, and alternative routes.

If the brain's obvious associations could be inhibited, then the brain would make new leaps of thought, reasoned Bernardi Luft and her team. They were quite sure that anyone could acquire a greater range of associations if their conventional associations were repressed. One year earlier, Bernardi Luft's team had attempted to do this by passing electricity through the dorsolateral prefrontal cortex, the center for executive function. And the result was overwhelmingly clear: 32 percent of the participants solved the problems they were given when they received the electric current, compared with only 5 percent of those who received the wrong type of current or none at all.

"We can improve very specific think-out-of-the-box [processes], but at the same time we decrease working memory processes," explains Bernardi Luft.

This electric method can ruin your concentration, but increase your creativity when it comes to basic problem-solving. It's not especially useful when you want to juggle a lot of ideas simultaneously, such as when you are working on a book and need to keep track of all the chapters—because then you need a well-functioning working memory. What Bernardi Luft's experiment showed was that creativity increased when the executive function decreased, something that also occurs when you are tired.

"If you are a morning person and you are working at night, your dorsolateral prefrontal cortex is going to be suboptimal. So you can potentially use that in your favor to try to work on specific problems that you are stuck on, [as] you have a little bit less cognitive control," she points out.

In her latest experiment on the temporal lobe, Bernardi Luft and her colleagues gave the test subjects a placebo, and yes, it showed that the temporal lobe could be one of the keys to creative thinking. It is, after all, part of the executive function: when the brain is put in alpha (wakeful rest), the doors to our obvious associations close, and the paths to our more unusual ones open. As soon as I read about this, I knew I had to give it a try; it sounded like a bona fide *queen killer*!

When I asked Bernardi Luft if she could repeat her temporal lobe experiment on me, she showed me the equipment, which was locked in a separate office.

The test she performed on me involved questions about language, logic, associations, and consequence thinking—the same questions her own test subjects had been given—and was divided into four parts.

Before turning the power on, we did a baseline test (one without electricity) which I actually found quite difficult, since much of it contained not only logical mathematics, but several pages where I had to combine three words with a fourth that was often only vaguely associative. English isn't my first language so it all felt pretty frustrating, but I got through it anyway. For the consequences test, I had to answer questions like "What might happen if humans were unable to speak?" or "What can be made from a brick?"—Guilford's test. This went very well and I had masses of ideas.

I then put on a rubber cap full of holes that were perfectly adapted for the electrodes, which were in turn attached to the right side of my head and forehead, using a little jelly to conduct the electricity into my brain. Bernardi Luft assured me that the current was very weak and told me to let her know if it became uncomfortable, in which case she would turn it down. Slowly the EEG locked in on my beta waves and moved down into alpha to draw my brain waves from beta down to alpha.

At first I didn't even notice it; I just thought the light was behaving a little strangely. It started flashing. Then I realized that it was flashing only when I used my right eye. The flashing was coming from my brain! The current was passing through my optic nerve, which made it look like the light was going on and off.

It also tickled, and I had a really weird pulsating sensation in my head. The whole world seemed to be vibrating slightly—the light, the colors, and me; it was as if my brain was quietly humming. It was uncomfortable, but I felt like I needed to hold it together and go outside my comfort zone, to endure a little electric current for the sake of creativity!

Strangely enough, the current made me worse at solving the language problems and the mathematical ones. My mind just went blank when I tried solving them, yet I managed to solve nearly all of them when I tackled them without the current. I was so distracted by the pulsing in my head. But the consequence test, where I had to suggest what might happen if the world's gravity were halved, went much better. I had a huge giggling fit, and a bunch of weird images and ideas popped into my head—although the suggestions I wrote down were almost incomprehensible, clearly only making sense to me. I suddenly felt quite tipsy and stupid, and I regretted having drunk nothing but canteen coffee all day.

"What we're doing here is opening our minds to different ideas," Bernardi Luft told me. "Statistically, there have been more good solutions to our tasks while using alpha wave oscillation than without, so we have seen that this affects creative thinking and causes people to see a larger number of, and more creative, solutions than normal," she said.

Bernardi Luft also found it strange that I'd been seemingly blocked by the current for some of the tasks. Perhaps I was already so creative that the current simply disrupted the strange pathways in my brain? Or maybe something weird happened when I crashed headfirst into the bridge? The scientist couldn't really say anything about that.

"What if I wanted to become more creative by using an electric current in my brain, through alpha stimulation of the temporal lobe—how would I do that?" I asked.

"You'd have to wear this cap on your head all the time," she replied. "Your brain is only being stimulated while you are connected to the electricity, and maybe up to an hour after you've been disconnected."

The prospect of walking around with a rubber hat permanently attached to my head didn't feel hugely tempting. The obvious benefit would, of course, be the stream of weird and original ideas I might get. And it would make me less interested in social conventions and used to humiliating myself. But a rubber cap with electrodes attached would be a fashion statement too far.

But first, I needed to find out more about what had happened to me, behind the right side of my forehead, the part of the head I'd smashed so hard. The temporal lobe connects the language system and the limbic system—which controls all your emotions and is therefore the control center when it comes to the art of writing; the right temporal lobe is also more associative than the left, and usually the least dominant. The temporal lobe at the right temple is responsible for face recognition and nonverbal perception. It is where we perceive music and images, and where we perceive space. The temporal lobes, along with the frontal lobe, are the areas that control our executive function, which in turn affects our concentration and focus. While the temporal lobes block associative noise, sounds, and visual impressions, helping the captain of the ship to hold the course, the frontal lobe is responsible for impulse control and planning, and these areas of the brain are responsible for your more advanced social skills. Damage to the temporal or frontal lobe can drastically affect and change your personality. While investigating how my accident may have affected me, I found, among other things, that damage to the temporal lobe can ruin your sense of humor (something I had experienced) and your sex drive, which can vanish completely. It can also make you very irritable. Damage to the dorsolateral prefrontal cortex can also lead to a loss of direction and willpower, and a reluctance to take the initiative for yourself or others—all quite central elements for social functioning and having a good life. The first time this kind of brain injury was seen to have an effect on someone's personality was almost two hundred years ago.

Just as many of the secrets of memory came to be revealed by the failed epilepsy operation on Henry Molaison, interest in researching the frontal lobe was also triggered by an accident. Phineas Gage, a foreman on the American railroad in the mid-nineteenth century, was blissfully unaware of his future importance to brain science as he led the blasting operation to clear the way for a new railway line south of the village of Cavendish in Vermont. One afternoon, on September 13, 1848, the charge they had set exploded by mistake, just as Gage was inadvertently looking over his shoulder. Out shot the tamping iron they were using to pack the dynamite into the hole, at lightning speed. The iron rod, 3.6 feet long and 1.3 inches wide, passed through Gage's mouth, then his skull, and out the top of his head, with all the force that had been calculated to split the rock face. The blast was so powerful that the rod was found eighty-two feet away, smeared with blood and brain. Despite also suffering severe burns to his hands and arms, Gage was then taken—still alive—to the doctor. Unbelievably, he was sitting up when the doctor arrived, and mumbled: "Doctor, here is business enough for you." After the doctor had cleaned his wounds and removed several spoonfuls of brain tissue that had been dripping from the hole in his head, Gage said there was no need for his friends to visit him. He would be back at work in a few days, he claimed, full of energy—the same energy I had been full of immediately after my cycling accident. However, ten weeks passed before Gage was in any condition to do anything but go home to his parents in New Hampshire, and he never worked as a foreman again, one possible reason being that nobody recognized him anymore—his temperament and disposition seemed to have changed completely.

Gage appeared to be incapable of making plans or carrying them out. He had also become incapable of handling social situations, something that eventually improved when he took a job as a stagecoach driver. It's quite likely his inner critic was completely destroyed, along with his executive function. After suffering increasingly frequent epileptic fits, he was forced to give up his job, and finally died from a huge seizure, twelve years after his accident.

But there are many reasons to question whether it really was that simple. Gage had, despite everything, survived an extremely serious injury, and had lost his job. It's not terribly strange that he didn't seem as balanced as he'd previously been. Also, his story wasn't published until eight years after his death, by John M. Harlow—the doctor that treated him who was a keen exponent of phrenology, the brain research of the time. Could it be that the doctor wanted to portray Gage in a way consistent with the theories he believed in?

The phrenologists were particularly interested in attributing different human characteristics to different parts of the brain, and believed that it was possible to recognize these characteristics externally, by studying the shape of a person's skull. This science would evolve into a useful tool for social Darwinists and Nazis throughout the following century. It was then considered possible to determine if a person belonged to a superior or inferior race: you just measured their skull to find out! Phrenology's reputation as a science was certainly diminished after the fall of Nazism at the end of World War II. The frontal lobe is not a phrenologist's garden from which you can pluck a flower and see a particular result. The brain does not consist of unique and separate modules that manage themselves.

"Like any simplifying idea, it is seductive. It offers the illusion that it can provide an instantly available explanation," says the Russian neuropsychologist Elkhonon Goldberg. Goldberg is a former pupil of the internationally renowned Soviet scientist

Alexander Luria and has written several books about the frontal lobe, that is now his specialist area. He describes the frontal lobe as a system, a network—an orchestra conductor.

Executive function is a network in the frontal and temporal lobes that creates interaction in the brain. It orchestrates language, spatial awareness, sound and imagery, motor experiences, memories, and imagination, and it does this to give us direction and focus and to enable us to come up with new solutions and stories. When I injured my head, it affected me in many different ways; I became more disorganized, unfocused, and uninhibited, all normal symptoms of an injury to this part of the brain. I also stopped cracking jokes, something that I felt affected my personality. These things, combined, made social interaction more difficult for me.

And losing my inner critic wasn't something I found liberating, as you may have expected; it was actually quite stifling. Humor occurs during close interaction with your surroundings, and is dependent on your mastery of reading social situations and having funny "aha" moments, but I was unable to do this immediately after the crash. I also suffered tinnitus, and my eyes were not functioning as they should have; I suddenly needed glasses in order to stabilize my twitching pupils (a fairly common but also fairly unknown effect of concussion, I was told when I requested expert help at Oslo University Hospital). And although my head was overflowing with ideas, they weren't terribly easy to implement while I lacked direction and concentration. Being totally free was actually quite restricting.

"No other cognitive loss comes close to the loss of executive function in the degree of devastation that it visits on one's mind and one's self. As we learn more about brain diseases, we are discovering that the frontal lobes are particularly affected by dementia, schizophrenia, traumatic head injury, ADHD, and a host of other disorders," Goldberg points out.

Executive function was first described by the scientist Alan Baddeley, who since the 1960s has researched working memory. In 1974, he presented a model for understanding this system of the brain, and it is still in use today. It all started when Baddeley, as a young scientist, was asked by the British postal service to develop the best possible postcode system. This led to him discovering that we cannot retain more than nine characters in our heads simultaneously, which is what became the basis for his ideal postcode system—and after that, his model for working memory.

Working memory consists of both our ability to perceive and organize sensory impressions and our ability to focus and concentrate, which is in turn connected to what enters our consciousness. In addition, it is where memories and ideas are played out to our inner eye. But what executive function actually *is* remains a mystery. Have we really put all our intelligence and willpower into this network and, by doing so, just moved the central mysteries of humanity there, in miniature? Why do we choose to focus on one thing versus something else—is this not simply the mysterious idea of the "personality"? In the fifth and sixth centuries, alchemists believed they could create a magical, miniature person, a so-called "homunculus." In the 1950s, brain scientists surpassed this idea when they proposed that all of the body's functions were mirrored in the brain, like a driver operating an excavator from inside its little glass box. But this might be the wrong way of looking at personality altogether, and executive function is perhaps more a result of many different functions, systems, and collaborations.

"Whether we will then be left with a single coordinated system that serves multiple functions, a true executive, or a cluster of largely autonomous control processes—an executive committee—remains to be seen," wrote Baddeley when he later tried to sum up executive function. For him, there were still many unsolved mysteries. When Alice stands at the bottom of the rabbit hole, in the little hallway surrounded by doors, why does she choose the door that leads her to Wonderland?

In his 1995 book *An Anthropologist on Mars*, the neurologist Oliver Sacks writes about "the last hippie," a man who lived as a monk at a Hare Krishna temple in the US.

After several years of worsening eyesight, the man had gone totally blind, and apparently lived in a perpetual state of religious happiness with a vague smile on his face. Being so detached from anything of value, he seemed quite holy, but the man was not quite as blessed as you may have thought—he had a benign tumor the size of a grapefruit in his frontal lobe. It was this that had robbed him of his eyesight and his interest in the outside world. For the rest of his life he was apathetic, with no connection to the world or his family; he was disengaged and unmotivated. The only thing he seemed interested in was music. He was certainly no longer bothered by social control or an overbearing inner critic, yet he was detached from his fellow humans and had somehow lost... direction. As mentioned earlier, what this direction consists of is fairly unknown. What we do know is what makes it disappear: damage to the frontal lobe.

The frontal lobe is the "youngest" part of the brain, in an evolutionary sense; it was the last to evolve and is largest in humans, compared with the size of our bodies. It manages not only our advanced social skills—in terms of language, shared culture and context, and our memories and ideas—but also our inner critics, our drive and concentration, our sense of social rules and inhibitions with all their pros and cons, and, yes, the mysterious consciousness.

Following my visit to the University of London, I now know that passing an electric current through my brain's right temporal lobe made me feel unusually spontaneous and a little drunk, and that it was while I was in that state that I had the most good ideas. So why use a rubber helmet and electricity to make myself feel tipsy? Why not just drink alcohol? And what is the connection between losing your inhibitions, creativity, madness, and intoxication?

The myth about drugs and creativity is still alive and well. It's hard to picture a creative role model from the 1800s without a glass in their hand. So perhaps there's something to it? By testing fifty people after they had drunk alcohol, and then testing the same people after they had been given a placebo (not alcohol), researchers in Dresden observed that the right frontal lobe became less active—and the subjects' inhibitions were reduced—after consuming alcohol, compared with when they did not consume alcohol. The test subjects' inhibitions were significantly reduced by the alcohol! But what the researchers also found was that those who lost their inhibitions fastest due to the alcohol were at a greater risk of becoming *addicted* to alcohol. The first thing that happens when you drink alcohol is you lose your inhibitions toward *drinking*. And from there you're off, not on your way to being more creative—most likely on your way to drinking more alcohol.

William B. Irvine, who has researched the lifestyles of artists and writers, says:

> Whenever we hear about great writers who were alcoholics, we need to ask two questions. First, when in the day did they do their drinking, while they were working or afterward? Lots of writers who drank did so after they had finished working; this is what Hemingway did. We would find the same thing, of course, if instead of looking at writers we looked at lawyers, brain surgeons, or factory workers. Second, at what stage of their career did they do their drinking? Many artists with reputations for being alcoholics didn't start their careers that way. In some cases, they were led to drink excessively by the pressures brought on by the literary success they enjoyed as young, sober artists.

As tempting as it is to suppress the control and inhibitions stemming from the right frontal lobe using alcohol, it's perhaps better in the long run to do what the celebrated Japanese author Haruki Murakami does. Murakami swears by a healthy lifestyle—eating mostly fish and vegetables, and going for a run every morning after getting up at 5 AM. He is extremely disciplined, and believes that living healthily assures him a long life as a writer—and he doesn't need drugs and partying to unleash his creativity. In his nonfiction book *What I Talk About When I Talk About Running*, he has this to say about the myth of the artist:

> A lot of people in Japan seem to hold the view that writing novels is an unhealthy activity, that novelists are somewhat degenerate and have to live hazardous lives in order to write. There's a widely held view that by living an unhealthy lifestyle a writer can remove himself from the profane world and attain a kind of purity that has artistic value. This idea has taken shape over a long period of time. Movies and TV dramas perpetuate this stereotypical—or, to put a positive spin on it, legendary—figure of the artist.

The myth of the intoxicated and liberated artist has roots stretching back to ancient Greece, when artists were inspired by semi-godlike muses and poetic madness, and took part in Dionysian rituals—Dionysus was the god of wine—that involved drunkenness and losing one's inhibitions. In the 1400s, things really took off when Marsilio Ficino launched his ideas about Saturn and melancholy, bringing the darkness of the mind into the mix. Then in the 1800s, we really cultivated the idea, when Goethe's moodiness and Byron's total lack of inhibitions were seen as leading mythical artistic forces—outsiders exploring the limits of society. Lord Byron even had a sexual relationship with his half sister; you don't get more exploring-the-limits than that.

"And at that same time, Mary Shelley wrote *Frankenstein*, which says it all, really," says Tracey Emin, referring to Byron's friend who wrote one of the most iconic novels of the nineteenth century. Mary Shelley was a highly disciplined woman, not someone who was engulfed in wild partying.

The artist Tracey Emin no longer drinks alcohol, after a recent battle with cancer. But even prior to that she had for many years kept her drinking to a minimum, for the sake of her brain. Without a well-functioning brain, she cannot make art.

"It's not being holier than thou, it's that I know what I'd rather be doing when I'm ninety-five. And it would be so stupid not to be able to do that because my brain had gone to mush ," she says.

Besides, some things are more potent than alcohol. "The intoxication of love is the most powerful thing!" she says.

The power of emotions—such as love and lack of love—is the one overriding force behind Emin's work, which frequently leads her to explore themes of sexual abuse and loneliness. One of her most important guiding lights is the Norwegian artist Edvard Munch, who in many ways helped shape the myth of the artist into what it is today. Emin discovered Munch as a teenager, when she stumbled upon a book containing one of his pictures and was instantly captivated. She had never seen anything like it before.

"I fell in love with Munch and his work the moment I saw it, and I feel the same way now over forty years later. I think he was very honest about himself," she says.

Munch's art is very self-exploratory. His *Self-Portrait With Bottles* from 1938 is echoed in Emin's *My Bed* from 1998; both pieces depict the excessive and, at the same time, shadowy, unglamorous aspects of each artist's life. Munch paints himself in front of a table full of empty bottles, while Emin exhibits her own bed, its dirty sheets covered in bodily fluids, surrounded by empty liquor bottles, condom wrappers, and personal effects.

"Exhibiting my bed wasn't an idea; it was an epiphany!" says Emin.

"Do you feel brave?" I ask.

"I feel brave making art and it scares me sometimes. When it comes to my depictions of rape, I don't choose it. It chooses me. It's a subject matter that comes up every few years and I just work with it when it does," she replies. (For people not familiar with Emin's work, some of her art is a response to her rape as a teen.)

The world of international art is permeated by conceptual art which is often overintellectualized, and, although creating art doesn't normally require physical strength, it has become a very male-dominated arena. Most female artists are overlooked and underpaid.

"I think my work was overlooked and not taken seriously for years, not just because of the materials I used, but because of my subject matter," she says.

Emin doesn't allow herself to be limited by themes or materials; her work consists of embroidered rugs, small sculptures, and paintings, and she recently finished a bronze statue that will stand prominently outside Oslo's new Munch Museum.

"Munch's mother died when he was very young," she explained to the British newspaper the *Guardian*. "So I want to give him a mother."

A thirty-foot-high mother, to be precise, upscaled from a small clay figurine. The sixteen-ton statue, complete with the artist's equally upscaled fingerprints, is inspired by her acute sense of vulnerability and, despite its colossal size, a sense of intimacy and emotional truth—a product of the forces that always guide her artistic compass.

"I never look to culture for inspiration, be it high or low. My work is about emotion and how I feel. I don't get a good idea; I get a feeling. Ideas grow from my mind and the way I think, from day to day. It's never a technical process. It's a feeling. An emotional feeling. Sometimes I don't even know what I'm going to do. I might see something that could be a device for helping me explain myself. Recently, I saw a thumbnail of Manet's *Baudelaire's Mistress*, and for a second I thought it looked like me," she says.

Emin also makes videos and installations, in which she often portrays herself. Through art she processes the rape she suffered, the abortions she experienced, and the loneliness, insomnia, and longing she feels.

In 1998, she filmed herself in the idyllic Norwegian coastal town of Åsgårdstrand, where Munch often spent his summers and painted many of his key works, including *The Girls on the Bridge* and *Melancholy*. In Emin's video we see her naked and curled up on Åsgårdstrand pier, while a piercing scream can be heard. This work, called *Homage to Edvard Munch and All My Dead Children*, echoes Munch's deeply emotional depictions of his sister dying from tuberculosis.

"I spent quite a lot of time alone as a child, too, so I think I was always introspective. And even when I was little, I liked creating things. I used to make doll's houses out of cardboard boxes," she has said.

She has always had a strong desire to express herself. But as well as producing an enormous and unstoppable flow of art, Emin is almost obsessed with expressing herself in writing.

"I don't think I can live without writing. It's a place to put my thoughts. I used to keep a diary but then I thought it was dishonest because I knew that one day someone might read it. Now I deliberately write for other people to read, on Instagram for example."

One aspect of the myth of the artist has, however, proven to be true: traveling, as Rolf Reber's research has shown. Travel offers us new perspectives and insights, and it is often when many "aha" moments occur. For Emin, traveling is central to both her art and her life.

"Everything is a journey. Every moment is a journey, and my art is, of course, about myself."

In the past, artists often belonged to the upper class and lived wild lives that involved trips to exotic locations, lots of alcohol, and raging emotions—things poor people couldn't always afford to have in those days. Now, however, most Westerners have access to all these things without them necessarily leading to creative output (you're quite welcome to check this yourself by going on a trip to Ibiza).

"In the 1800s, you're talking about a culture with extreme social tensions; the poor and the bourgeoisie lived in entirely different worlds. Many artists visited the lower tiers of society to gather inspiration, and they idealized madness and intoxication. Our modern ideals are drawn largely from there," says the psychologist Peder Kjøs.

Throughout the 1800s, the myth of the artist was kneaded and shaped into what we recognize today: Goethe, Schubert, Van Gogh, Munch, and Nietzsche all helped create the impression that an artist needed to experience intoxication and excess, melancholy, and madness. The Greek myths, the romance of drugs, and Ficino's dark melancholia were all combined to form the cliché of the suffering, excessive (and male) artist.

"Artists have a kind of experience at the frontier which they share with the rest of us. They dare to do something the rest of us don't. They exist in the unknown territory between the normal and the strange; they look beyond the curtain we others don't dare to lift. They are people with religious experience, researchers, artists, and I'm glad there are people exploring that particular area. It's something we all admire, of course. I genuinely believe in the myth of the artist. Some people are more in tune with the universe; they see something bigger and greater," says Kjøs.

In tune with the universe and unrestrained by societal rules and norms—sure, that would be nice; then I wouldn't get so much trouble from my inner critic. But the idea that madness is a part of what it means to work creatively—that mental illness liberates you from social conventions and restrictions—belongs to the myth of the artist. And unfortunately, the reality is more like my cycling injury—*not being yourself* is of no advantage when you want to be creative. Mental illness is more of a hindrance than a help if you want to stretch the boundaries of art.

"Many people think that schizophrenics are more creative because they 'think outside the box,' but only 2 percent of those patients are involved in anything creative. It is also difficult to appreciate art created by psychotic people, because we expect some kind of coherence and consistency in a creative work. If a creative work is too inconsistent and strange, it will miss the target; people won't be able to understand or appreciate it. Madness is not the road to creativity. There may be a large number of people with bipolar disorder working in the creative industries, but there is no proven advantage that psychiatric illness is beneficial to creativity. Normally, they will be too far away from what most people are able to understand," says Professor Bhattacharya.

According to one report, authors are ten times more likely to suffer from bipolar disorder than the rest of the population. The figure is even higher for poets: forty times greater than average. This could be one of the reasons why Aristotle described artists and thinkers as melancholics: he perhaps didn't realize they were bipolar, and noticed only the depression. In any case, none of this proves that you need to be bipolar to be a good poet! Unfortunately, the connection between madness and genius has proven to be a persistent myth.

Vincent van Gogh has, in our time, become the epitome of the ultimate artistic genius—one who "cracked." Van Gogh lived a fairly isolated life—in fact, his only close relationship was with his brother Theo. He began painting at the age of twenty-eight, and, in the space of only ten years, produced more than two thousand pieces of art. After a string of bizarre incidents, one of which involved him cutting off part of his own ear, he was committed to an asylum. Prior to that, he had binged on absinthe, chain-smoked, and drunk far too much coffee, while also not eating a great deal. It is also believed, absurdly enough, that he gnawed on his own canvases—and, of course, paint isn't especially good for you. (At the time, it contained all kinds of lethal toxins like arsenic, which was regularly used in green paint to make the color appear more lush and lifelike.)

A lot of research has been done to establish what kind of illness Vincent van Gogh actually had, and which eventually made him take his own life. Diagnosing people from another time and place is, of course, slightly problematic. Nevertheless, many people have tried solving this mystery and many theories have been presented: bipolar disorder (of course), along with epilepsy, lead poisoning (from the paint coloring), hallucinations (from the absinthe), tinnitus (as a result of Ménière's disease), Geschwind syndrome, and persistent heatstroke (after standing in the sun for too long, painting), to name a few. But vague diagnoses like these are rarely correct. What is quite clear, however, is that it wasn't the artist's life or the mental illness that was the key to Van Gogh's creative talent. These were more likely things that he had to fight against in order to be able to create his art. When his illness got the better of him, it put a stop to his enormous turnover of work—and in the end, it took his life. In his lifetime, he was never a sought-after artist.

"With regards my chances of sale, look here, they are certainly not much but still I do have a beginning. At the present moment I have found four dealers who have exhibited studies of mine. And I have exchanged studies with many artists. Now the prices are 50 francs. Certainly not much but—as far as I can see one must sell cheap to rise and even at costing price," wrote Van Gogh.

Divergent thinking involves having many thoughts that are well beyond the regular societal norms and associations—but not too far out either, as Van Gogh's perhaps were. This could be why, of the almost nine hundred paintings he made, he failed to sell more than one during his lifetime. Granted, he was hugely successful after his death, but most of us would rather have success while we're alive, and would certainly want to feel like we're understood by the people we look up to and respect.

More recently, we have found other ways of exploring the fringes. One of the things Berkeley professor Michael Pollan looks at in his book *How to Change Your Mind* is how drugs actually work, and how LSD can make the process of dying easier for end-stage cancer patients.

The connection between drugs and creativity has been nurtured for centuries. Take Sherlock Holmes and his cocaine addiction, for example, or the fabulously colorful poem "Kubla Khan" by the Romantic poet Samuel Taylor Coleridge, which was obviously written after an opium dream. Many interpret *Alice's Adventures in Wonderland* as one long drug fantasy, a presumption supported by the hookah-smoking caterpillar that appears in the book, and by Alice eating a mushroom to make her change size.

So-called mind-expanding substances have been extremely popular since the 1950s. One of the most important defenses of psychedelics can be found in Aldous Huxley's 1954 book *The Doors of Perception,* a book about the author's experiences while under the influence of mescaline. The most important connection between LSD and creativity was perhaps established in the 1960s in Los Angeles and New York, where therapists began giving the drug to their patients as a part of their treatment. Famous patients who have vouched for this treatment include Jack Nicholson, Anaïs Nin, Stanley Kubrick, André Previn, and Cary Grant.

"A man is a better actor without ego, because he has truth in him. Now I cannot behave untruthfully toward anyone, and certainly not to myself," said Grant in an interview he gave following treatment. Demand for LSD exploded as a result.

The latest thing today is microdosing with LSD, something that is popular right now in Silicon Valley and is a growing trend. One of the great inspirations for modern tech entrepreneurship is the late Apple CEO Steve Jobs, who spoke warmly about his experiences on LSD and of meditating in India; these experiences were considered to be his creative dynamite. These days, LSD consumption is more controlled than in the 1970s, but with these microscopic doses, young technology talents hope to increase their creativity, while avoiding the distracting, powerful drug hallucinations that would normally have made performing in the workplace extremely difficult. Instead, their goal is to facilitate unconventional ideas.

A brand-new study of LSD microdosing, carried out on ninety-eight subjects in Australia, isn't quite as positive. It suggests that LSD does not have the creative effect the users are hoping for, and the researchers behind the study have pointed out that a controlled study would be necessary before a proper conclusion could be made. (A controlled study involves dividing test subjects into two groups and then giving each a medicine—one a placebo, the other a real substance—under identical conditions. Setting up a controlled study of an illegal drug using unwitting test subjects would therefore be very difficult from an ethical point of view.) At the same time, a new study in Toronto has shown LSD to cause an increase of creative activity, although this study was also performed without controls. I could find only two studies on LSD and creativity on the research database PubMed, and both used people who had been recruited because they had reported prior use of psychedelic substances.

The problem with studies of this substance is that the test subjects need to sign up for the trials voluntarily—which means the participants are especially interested in the substance, and have perhaps taken it quite a few times prior to volunteering for the study. The results can therefore be affected by the volunteers' own desire for the test to have an effect.

"So, should I take LSD if I want to be more creative?" I ask Michael Pollan.

"I'm not sure if you should take drugs. There's a lot of work yet to be done on this, and I'm hoping some of the new generation of researchers will work on it. There's good reason to believe—based on both the neuroscience and anecdotal reports—that psychedelics should improve divergent thinking and pattern recognition, as well as afford the sort of fresh perspectives that often accompany creativity. But this hasn't been proven to anyone's satisfaction," says Pollan.

Considering what I know about studies of creativity and psychedelia, I think that exposing my poor, battered brain to LSD would be a little too risky. It's also against the law. Besides, I've got no idea where to buy it. I could try marijuana of course, which according to *Dope Magazine* is meant to be very good for creativity. I can't find any research proving that hash smoking is a good idea, although I do know where to buy it (my local corner shop), not that it makes me more tempted; maybe I'm wrong, but I doubt the scary-looking guys sitting behind the vandalized shop door are at all familiar with the latest research. Anyway, a test carried out in Israel using a placebo-LSD showed that people became more creative by simply *thinking* they had taken something mind expanding.

At this moment in time, I can think of only one practitioner of the arts who maybe—well, perhaps definitely—takes drugs. For Norwegians he is a familiar voice on the radio, a TV personality, a scriptwriter, and a musician. He founded and folded a record label that released one hundred records in the space of two years, and he has made a number of popular podcasts. In 2001, he sat in a display window on Oslo's busiest shopping street for a whole week and ate junk food; it was both a performance and a PR stunt, which he called "Decline," and the subsequent TV show was nominated for one of Norway's most coveted awards, Gullruten. He wrote a critically acclaimed book in 2009 about people whose funerals take place with no friends or family present, *On Behalf of Friends*. And he wrote a series of articles about the trial of mass murderer Anders Behring Breivik, for the newspaper *Morgenbladet*. His output is very often surprising and different, and he does not seem restrained by any inner critic.

"I know a lot of heavy drug users. And I'm sure they see themselves as very creative, although perhaps not in a way that anyone else would appreciate," says Kristopher Schau with a grin.

Schau is actually not a drug user himself. He did, however, eat his own mole on live TV, and once dismantled an entire hotel room for no particular reason. But Schau's activities on the fringes are more likely the result of his own distinct attitude toward life and the people around him.

"I don't really care what people think or how many of them listen or watch what I do," he says, often using the words "an adult" when referring to people he has worked with.

You could say that he has an amazingly disciplined way of seeing himself as a child. His alarm clock rings at eight, which is followed by a totally ironclad morning routine: He turns off the alarm and turns on his PC in one movement, then switches on the coffee machine, then goes to the bathroom. Then he takes his cup of coffee and sits in front of his PC to read through and make adjustments to whatever it was he wrote the previous night. He is a bit like a modern-day Ibsen with regard to having order and routines around things; Ibsen was known for living a very routine artist's life, and in his later years you could have set your clock by the playwright's strolls along Karl Johans gate, which he took at precisely the same time every day. Kristopher Schau's routines are like that.

"It goes really slowly. But I just write without thinking; it's like my head's completely different in the morning. I'm so tired I can't even speak, but strangely enough I always manage to do it," he says.

This routine seems to be incredibly productive for Schau, who works tirelessly on all sorts of projects. "I've forgotten most of what I've done. I never think about it afterward, other than remembering it as a good story. I just think about what I'm doing *next*. I like making things, whether it gets an audience or not. It just feels satisfying; that's the only way I can explain it. And it seems like the more I do, the more ideas I get! Because what's to lose? As long as I can pay the rent and put food on the table, not a lot can go wrong. It's no big deal," he says with a shrug.

At the time of writing, Schau and his band The Dogs are preparing to release a new album, but he is more interested in working on all the ideas than fiddling with the fine-tuning. He has also performed with six other bands, including Hurra Torpedo and Gartnerlosjen.

"When Gartnerlosjen started, the huge trashing we got from critics was actually a stroke of luck. Anything positive I hear now just feels like a bonus," he says.

Primarily a musician, Schau loves playing live. He tells me how The Dogs once performed a daytime gig to two eight-year-olds and two fourteen-year-olds in Bodø, a small town in the Arctic Circle. With the four children climbing on a sofa in front of the stage, Schau gave it everything he had, smashing the mic stand between his legs so hard it made him throw up. He found out later that he had sprained his wrist.

In hindsight, he now says the best concert he ever played was a 5 AM sound check for a TV appearance on *Good Morning Norway*.

"We didn't think; we just played, feeling the music. There was no audience obviously, just a bunch of sleepy camera operators, and we were extremely tired as well. But it went amazingly well!"

It was while writing a TV script that Schau learned his most important creative lesson, and why he should never rely on claims about "what most people want," or what conventions and rules he should follow.

"*Kill your darlings*—that has to be the stupidest thing I ever heard! If there's anything you shouldn't kill, it's your *darlings*! We sat writing a script for a TV series about three couples who were in therapy, and there was one character we really loved. And then the TV company said he just wasn't at all believable and they wanted us to write him out of the series. But we believed in this character, so we did the opposite: we made an entire series about him instead, *Dag*, and it became a huge success. *Never* kill your darlings. If you're ever in any doubt, get rid of everything else! I firmly believe that's true, that if I love something, then there will be someone else out there who loves it too," says Schau.

Something tells me I need to be more like Kristopher Schau. There's no doubt that listening to someone so totally unconcerned with what people think—and who is so rarely controlled by an inner critic—is very inspiring. I think I may have found my muse! Picasso had Dora Maar; Leonard Cohen had Marianne Ihlen. The Greeks had nine graceful women who sent them ideas from the realm of the gods. My muse is Kristopher Schau, a black-clad beanpole of a guy with a straggly beard. I'm going to be more like him. If there's only one interested reader out there, that's good enough; if you are the only person reading what I've written now, then great. If Kristopher Schau can perform to a bunch of eight-year-olds in Bodø and rock so hard he sprains his wrist and destroys his testicles, then I too can give it all I've got—for you, dear reader, since you've read this far and not given up. I won't be injuring my testicles, but I'll certainly try and sprain my wrist for you. Because now I know how to get rid of my inner critic. I have to do something radical. Kristopher Schau's untamed creativity has made me think that I should practice being more spontaneous.

Maybe it will help me beat the Queen of Hearts to the punch if I just do something downright absurd? It must be possible to run from a tiny little queen!

Keith Johnstone has fought for spontaneity most of his adult life. It was Johnstone who founded improv theater in England in the 1960s, when he was the director of the Royal Court Theatre in London and already deeply suspicious of the authorities. Setting up a theater in the British capital at the time required approval from the censors, something Keith cared very little for. But long before that, he had an aversion to anyone who wanted him to do things he didn't want to do. He realized from day one that he hated school, and was furious with his parents at their well-meaning attempts to get him to eat meat. So he became a vegetarian as a child, and as a director he devised games to use in his improv theater. These games diverged from the British postwar values of being sensible, in control, and well behaved. Johnstone's theater was a place where both the idea of authority and logical thinking were deprioritized. He was inspired by the wrestling matches he watched in England during the 1960s, bouts that switched from performance to play, then to sport and entertainment.

Paradoxically, Johnstone, now eighty-six, has ended up being an authority himself, and continues to hold workshops all over the world despite being well into retirement age. At these workshops, he teaches his students to *not* be good, to *not* try, to *not* concentrate—to *not* think inside the box. The most important rule for novices trying out improv theater is to say "yes, and"—in other words, meeting all your challenges with a yes, then building on whatever fantasy you have been given.

"I was consumed by the idea that what perceives and constructs the universe is a person's brain. Remove the inner critic, and we can all become geniuses," says Johnstone.

"I'm deeply suspicious of religion and all authority, and the school system. I think we can learn anything if it is a game or through play. My games will make you play out your creativity. I laughed so much during the time that I came up with improv theater; I laughed from morning till night," he says.

Improv is now a staple at all drama schools, and you'll see many of Johnstone's games on the TV show *Whose Line Is It Anyway?*, where well-known actors challenge each other through various improvisational exercises. And since Johnstone's methods are now used in the advertising business, improv has spread to the self-help industry. One of the hippest things you can do in London must be to go to an improv course. Not with the goal of becoming an actor or entertainer, but because improv courses do something to you—you learn something about yourself.

"The big problem for many actors is that they want to be the best; they want to shine. They want to be the alpha male onstage. They don't want to change. They want the stardom, not the real transformation of character. But that's not interesting for anyone to watch. Improv is not about personal gain; everybody wins if it goes well—the audience wins. And for that to happen, you have to follow your intuition," says Johnstone.

Being good and showing off is simply a way of generating applause; it rarely involves taking risks. However, exploring your creativity without knowing where you're going is genuinely scary. It can change you. It causes something new to happen, something neither you, nor your inner critic, nor your audience has ever seen before. In such a process, it's important to be able to handle doing something wrong, because if you're going to submit yourself to something totally unknown, you need to let go of everything you already know.

After talking to Johnstone, I felt inspired. Off with *her* head, I thought, while thinking about the Queen of Hearts. Now I'm going to do something that will really shock her, yes, and put her completely out of action.

"I'm going to learn how to make a fool of myself," I said to my husband, laughing smugly after signing up for an improv course.

So one Saturday morning in February, I found myself on my way to a locale in east Oslo for my first session. Eight other expectant novices stood waiting outside when I arrived.

"If you ask a bunch of five-year-olds if they can sing and dance, every single one of them will say yes. But adults will answer very differently, because something happens as we grow up. I hold courses where I give you permission to trust and follow your own impulses. So the question is: When can you give *yourself* permission?" says the course leader, Terje Brevik, before we get started.

In 2012, Brevik was working as an IT engineer, and had planned on taking just one improv course. Now his life has changed forever. After numerous courses and trips to the US, Canada, and Europe, and obtaining a degree in drama and theater communication, he started Tøyen Impro, which offers improv theater courses to both business and private individuals.

"I still don't know much about the future, but it's not quite as scary anymore. As long as I'm going in the direction of doing something I enjoy, my chances of hitting the right spot increase, and improv theater is what I want to do. So far, I've taken the liberty of building up Norway's biggest improv scene, I started the Oslo Impro Festival, Tøyen Impro, and I've also performed and taught improv in the US, Germany, Switzerland, Latvia, and Russia. And yes, I do wrestle with my inner critic. Obviously, the inner critic isn't *all* negative. It's the impulse that prevents me from doing things that are obviously stupid, like jumping off a cliff. Yet it's also what stops me from approaching the pretty woman at the bar. I wonder if we are the ones most responsible for limiting our options?" says Brevik.

Brevik leads all nine of us through a range of exercises, where we learn to create stories together, or we give imaginary gifts to each other. Throughout the course, he often underscores something improv's founder Keith Johnstone talked to me about: Don't be good. Don't be funny. Just try and make the others look good; listen out for what they want. During improv lessons, two things seem to be crucial: feeling safe and being fast. We get very little time to complete the exercises, and with so little time to perfect anything, there's no time for conferring with your inner critic. "OFF WITH..." she cries, but doesn't quite manage to say "HEAD" before the next exercise begins. In fact, it's extremely quiet from her corner for most of the two-day course.

"As children, we don't protect ourselves in the same way that we do as adults, nor do we have the same ego or intellectual control over our bodies. Until they are four or five years old, children all over the world have a fearless attitude toward exploration and a healthy skepticism of the student-teacher relationship—something which will eventually and unfortunately suppress their inspiration and creativity," says improv coach Shawn Kinley.

Kinley is a former student of Keith Johnstone who also worked with him for twenty years. Nowadays, he travels the world nine months of the year teaching improv. The other three months are spent at home in Canada.

"When I teach improv at workshops for adults, I work on balancing their egos. I try to encourage the sense of self in those who minimize themselves to a detrimental point. At the same time, I try to 'readjust' and offer a healthier perspective to those who are blinded by their own sense of worth, to pacify their rational thinking and allow their automatic cognition to be more effective. You don't think consciously about what you're doing when you're riding a bike. I use games to create positive stress, to put us in touch with our bodies and our spontaneity, which we see as creativity. There are no right answers for the tasks I set; it's more about letting go of the ego."

What Kinley talks about reminds me a little of meditation and Buddhism, and for good reason: he has immersed himself in Zen and Taoism, where the goal is to let go of your consciousness. He refers to books like *Zen in the Art of Archery* and talks about how you should be the archer, the bowstring, *and* the arrow—to release yourself, become one with the world. But what does this have to do with game playing and doing things fast, two exercises that are clearly important to improv theater?

"If you say, 'Just be creative,' you end up with a vacuum. Saying that everything is possible is too open. You have to give people a specific task. Many of the exercises in improv theater are there to simply open you up for creativity. Many people think that improv is about saying 'yes' to every challenge, but it can rapidly evolve into a cult of the yes." ("Yes, and" has become a known formula in the PR industry.) "Really, it is about optimal communication, about sharing. When I'm teaching, I hope to learn something myself. It's part of the process," he says.

"You cannot erase the self completely, nor should you try to," Kinley emphasizes. "You have to *want* something, to fight for your idea. But your ego also needs to be small enough for you to be able to let go of the internal and external distractions blinding you from what is more relevant in the moment, to pass your idea to others and let them take it further. It's something in between, a balance. Create something, discard it... make something else," he says.

"Maybe you should try meditation?"

It sounds like a good idea. And I know it can work: in 1982, the now world-renowned professor Toril Moi was a regular practitioner of so-called yoga nidra meditation. It changed her life.

"Back then I participated in a yoga class that lasted three hours. We started with physical exercises and moved on to breathing and meditation. Then we finished with fifteen minutes of yoga nidra, a form of deep relaxation. Every week, before the yoga nidra began, the class leader would ask us to think about something we really wanted, deep down. We couldn't mention it out loud, or share it with anyone else; we just had to think it. But my mind was blank. Week after week, I just couldn't come up with anything. But one evening, while I was lying on the floor, something came to me: I want to write a book," she says.

Moi, who is based in the US and Norway, is now a highly respected and award-winning author and thinker—the fruits of an idea she had in 1982, after a session of deep relaxation. I decide to find someone else who knows a lot about meditation and relaxing.

Before Siw Aduvill started doing yoga, she had a dazzling career as a juggler and circus artist. Educated in Sweden and Britain, she became Norway's first female circus director. The work poured in, and she was constantly busy. She juggled on huge theater stages and filled the most vibrant modern circus tents with life and imagination. But all that is now in the past. Today, Aduvill works as a yoga teacher and has written two books: one about yoga and one about the art of resting. After many years of hard circus work, she has changed direction and now focuses on what it actually means to rest—because when you rest, something happens to your body and mind that is quite amazing, something that has a significant effect on your creativity.

"Circus school was extremely demanding. We normally worked twelve-to-fourteen-hour days, and when I'd completed my education, I continued working long hours. I was very diligent and followed through on every project I started, one being a circus school for children—I was bad at delegating, yes. But mainly, it just didn't seem possible that I could get burned out from doing something I loved, and I loved working with the circus! The creativity, the color, the traveling—it was everything I'd dreamed about while growing up in a suburban apartment," says Aduvill.

The colors and the juggling balls completely took over Aduvill's life, until her body finally said stop. It was then she learned how to meditate and do yoga.

"When I lead a yoga nidra class, I try to bring you down to a level between wakefulness and sleeping. Many people fall asleep during this type of meditation, but what I'm trying to do is make your brain waves go from beta to alpha, so that you relax and lose your inhibitions," says Aduvill.

This was precisely what I had done in London; I had lowered my brain waves to alpha—although lying motionless while drifting between sleeping and wakefulness, as you would during yoga nidra, is, of course, more comfortable than connecting your brain to an electric current.

"In a way, what I achieve here is similar to what happens when you are worn out. Which is how we were at circus school: we worked so hard we couldn't think anymore; we didn't have the energy for an inner critic. While improv theater gives you a deadline preventing you from being able to think, I go the opposite way. I calm the system down, a bit like when you drink a glass of wine," she says.

With yoga nidra you nurture your self-love, which is something that will make the Queen of Hearts speak a little quieter. It will also make you less inhibited, and more able to make unusual leaps of thought. In other words, there are ways of freeing yourself from your inner critic. Either you can run extremely fast, making it impossible for the queen to catch you, or you can give up the fight. Speed or silence. Escape, play, drink. Or work on your self-love. There are many ways to reach your goal, and they are all quite different.

"Our nervous system has two fundamental components for responding to this: the parasympathetic system and the sympathetic system. And these control our behavior in many ways," explains Aduvill.

"When your sympathetic system is activated, you are in fight-or-flight mode; everything you do is driven by stress and a need to survive. But with your parasympathetic system activated, you are in a far more open state of mind, where you are creative; you cultivate friendships, your sense of belonging, sexual relationships, and general well-being. While *sympathicus* is easily triggered and fairly persistent, *parasympathicus* is the more shy-and-silent type."

The two systems are mutually exclusive. One puts the body in panic mode: Your pulse rate increases, your blood pressure goes up, your heart beats quicker. Your stress hormones, like adrenaline, noradrenaline, and cortisol, get pumped through your body, enabling you to react faster. Your white blood cell count decreases, and your immune system worsens (put on hold, to be precise). Your digestive system changes too (in extreme cases, you can instantly lose control of your bodily functions); you sleep worse and eat less, and your sex drive either worsens or vanishes completely. Panic mode also makes your pupils dilate, enabling you to take in all your surroundings. Not terribly strange if you think you're about to get eaten by a lion—which is, after all, why we have this particular stress response. As we saw in the study of fast and slow thinking and "aha" moments, we prefer things we are familiar with. We are more interested in repetition and prefer to have confirmation of what we already know, when we are in stress mode. When we think aggressively and instrumentally, we get tunnel vision—which is nice if you're running from a lion, but is the opposite of being creative. When you're in *sympathicus*, you are not actually sympathetic. This state of relative blindness and targeted behavior conflicts with the whole essence of creativity: openness, flexibility, and curiosity (no one is curious when they are really stressed)—as well as being playful, imaginative, tuned in, and present both for yourself and for others.

Creativity belongs to the parasympathetic network, where you can cultivate love, friendship, and non-targeted behavior in peace and quiet. But if you continually attack yourself with an inner critic, your sympathetic system will activate, and you'll be driven into a stress response. And by doing so, you'll have destroyed your own creativity.

If I could just spend more time in parasympathetic mode, it would not only benefit my creativity; it would also be good for my overall welfare. In *parasympathicus*, I can demonstrate self-love and consideration for myself, and leave my inner critic with very little to work with. So, is it possible for me to control my parasympathetic network?

"Of course it is," Wasim Zahid, a cardiologist, assures me.

"Even though the parasympathetic and the sympathetic system don't really allow themselves to be controlled by our will, they do so anyway. If I measure a patient's blood pressure and the results are a bit strange, I'll ask them to go into the waiting room and calm down a little; I'll then take the reading again and the result will be different. So I'm absolutely sure it's possible to control this, despite it being called the autonomic nervous system," he says.

The important "switch" for the parasympathetic system is the vagus nerve (*vagus* is Latin for "wandering"), which extends from the brain via the vocal cords and through the body to the intestines. It is connected to our throats and our breathing, and can be stimulated by singing and laughter—and by yoga and deep breathing. The breathing techniques used in yoga are meant to put you in *parasympathicus*, even when you're in the middle of a demanding physical exercise. After twelve years of yoga, I know how to make myself relax, even when balancing on one leg, or in the bridge position. Finally, I'm starting to understand why I always write best after starting the day with an hour of yoga. Considering what I now know about stress, breathing, and creativity, it's tempting to say: Breathe deeply, and be inspired!

This is precisely what happens when comedian Beth Lapides does yoga and meditates, two activities that are essential components in all her creative work.

"I work with not knowing, and self-acceptance, which is usually the opposite of what the world demands of us. We're expected to know and have opinions about everything. Being intelligent and not having ready-made answers is clearly undervalued in our culture. I meditate every day and I do yoga, but you can just as easily achieve the same thing by going for a walk. There's something about putting one foot in front of the other. When I do yoga, I learn through movement. There are so many life principles that can be acquired this way," she says.

For almost thirty years, Lapides has been running her unique comedy night, UnCabaret, where comedians tell true stories from their lives. She has been called "the godmother of alternative comedy" by the *LA Times*, and has also done everything from acting in the comedy series *Sex and the City* to teaching creativity, hosting shows, and working as a producer.

"Yoga is about facing your fear, and about the marriage of flexibility and strength. I also learned a lot from the serenity prayer at AA, about accepting what you cannot change, and understanding what you can change. As a child I wanted to change the world, but as I matured as an artist I realized the world is going to change anyway—but I can still contribute," she says.

To want something, but not want it too much, to let go, is an interplay. Dr. Audun Myskja believes the two systems are constantly interacting, and that there are positive forms of *sympathicus*. Obviously, too much relaxing doesn't generate the energy we need to create something. Some kind of positive stress (like a white rabbit that's running away from you) is, as a rule, quite necessary in a creative process, somewhere halfway between *parasympathicus* and *sympathicus*.

"It's more of a constant interaction, similar to how Eastern medicine describes the dance between yin and yang," Myskja writes about the two divisions of the autonomic nervous system in his book *Breathe*, where he points out that a positive *sympathicus* can *also* lead to creativity.

"*Sympathicus* during phases of productivity is linked to enthusiasm, commitment, play, and interest, just as children are drawn to one after the other," he writes. Myskja teaches readers his different breathing techniques to help them come out of the negative *sympathicus*, where you are in fight-or-flight mode. As a doctor, he happily gives his patients breathing exercises as part of their treatment.

If breathing techniques aren't your thing, you can still try controlling your circumstances. Do you constantly put yourself in situations where you are facing a "lion"—be it an internal or external threat? Do you constantly bump into your furious inner critic, or external critics that make you frightened or angry? Because it is quite possible to facilitate a parasympathetic life for yourself—by constructing a mental fortress (and moat) around yourself—where you can live safely and creatively with people who wish you well. *Parasympathicus* is about celebrating community and friendship, not the lonely genius. Note that researchers found that women are more sensitive to criticism and require more external confirmation than men. Studies confirm that this occurs both at the elementary school level and among adults, but it is hard to determine whether this is biological or cultural. It suggests that togetherness is even more important for women who want to be creative than for men. And having an inner critic isn't quite as bad when you're surrounded by a good community, where you look after each other and where your ideas will be more readily accepted.

These communities are also the latest thing in brain research. At the Queen Mary University of London, for example, Caroline Di Bernardi Luft is busy looking at how brains work together. By connecting an EEG to the heads of two test subjects, Bernardi Luft has found that when people look deep into each other's eyes, their brain waves begin to harmonize. If the two people are friends or in a relationship, this harmonization occurs faster. She is also looking at how people work together to solve creative tasks.

"I think this is the future. We have to look more into how we are connected to each other, rather than at our individual brains. Neuroscience has been far too focused on zooming in on the singular brain," she says.

I'm quite touched by how brain researchers and creativity experts want to examine friendship and togetherness on a brainwave level. We're all connected, of course, shaped and restricted by our perceptions of each other; we shout "Off with her head!" now and then, but normally we support and take care of each other. We are woven together through culture and language. We are incredibly dependent on each other. As a species, we humans spend the most time raising a child (relative to our life cycle) and have succeeded in dominating the earth, not through being seven billion lonely geniuses, but through being seven billion people who collaborate. As my plane takes off into the darkness above Gatwick and over the sparkling lights of the city spread out below, it becomes clear to me: London is a strangely symbiotic fusion of millions of people and animals—irritating each other, exploiting each other, jostling and quarreling with each other—but who first and foremost take care of each other, and who exchange, share, and celebrate ideas. And now I'm on my way to Oslo, a city equally born of cooperation and symbiosis. For a moment I feel light and happy, and I suddenly think about what Aristotle says in the *Nicomachean Ethics*: "Without friends, no one would choose to live, though he had all other goods."

I run my fingers over my scalp, as though searching for the rubber cap and electrodes used in the test, and pick some jelly out of my hair. Maybe it's just the effects of the alpha-wave stimulation, or the semi-weightless feeling brought on by the plane, but for the first time in months I think: I'm going to do this. I'm going to write a fairly original book about creativity and send it out into the world, to one reader, to you.

3 | Playing Croquet With the Queen of Hearts

OR: WHAT IS IT YOU ARE ACTUALLY DOING WHEN YOU MAKE SOMETHING?

"When I used to read fairy tales,
I fancied that kind of thing never happened,
and now here I am in the middle of one! There ought
to be a book written about me, that there ought!"

"TO LIVE IS—to fight possession of heart and brain by the troll. To write is—to sit in session judging one's very soul," wrote Henrik Ibsen in 1878, sixteen years after *Alice's Adventures in Wonderland* first appeared. Ibsen makes the writing process sound quite uncomfortable, like an encounter with an almighty inner critic, or like a psychologist appointment on steroids. Lewis Carroll, on the other hand, was driven to write *Alice's Adventures in Wonderland* by friendly nagging. He would often take the three young Liddell sisters on boat trips, through Oxford and up to Nuneham Courtenay or Godstow, and one July day in 1862, Carroll (whose real name was Charles Lutwidge Dodgson) came up with a new fairy tale for them.

"Here from all three came the old petition of 'Tell us a story,' and so began the ever-delightful tale. Sometimes to tease us—and perhaps being really tired—Mr. Dodgson would stop suddenly and say, 'And that's all till next time.' 'Oh, but it is next time,' would be the exclamation from all three," wrote Alice Liddell, looking back, as an adult, on the events that led to the fairy tale's creation.

Ibsen and Carroll sound like they used quite different creative methods, which, of course, produced quite different types of literature. Yet creating something new must involve *some* common factors. But what are they? What happens in your brain when you work creatively? What are you actually doing? You, Alice, have the Queen of Hearts in an iron grip between your two fingers, while the Cheshire Cat smiles at you and the White Rabbit races ahead: What's actually going on, when something that wasn't there before . . . suddenly is?

"People hold artists in such high esteem, but we are all writers when we are sleeping. You could perhaps compare all creative activity to having a waking dream. When you're dreaming, you experience things and are surprised by what you see, despite it coming from your own mind. You don't have total control, and you don't know everything about what's going to happen. Making art is like having a lucid dream, where you have very little control over your own dreamworld. Many people have compared reading and writing with dreaming," says Bruno Laeng, a psychology professor at the University of Oslo who works with creativity.

"But it's important to remember that what separates art from dreams is the artist's ability to put their fantasies into an artistic format. And to do this you have to practice, practice, practice—ten thousand hours or more, if you're going to master an art form," he says.

To create new patterns, you need to be familiar with many of the old ones. An artist will create patterns within something chaotic. Scientists will do the same, as will architects, carpenters, and engineers. Working creatively means creating order from chaos, making totally new and interesting patterns where previously there had been a mess, or a totally different pattern—or just a stone. Michelangelo said that "every block of stone has a statue inside it and it is the task of the sculptor to discover it." This is an ancient way of viewing creation.

The Iranian Manichaeans described their god as one that organized matter—not one that created something from nothing, as the Hebrew God does in the Bible.

"Sinful, dark Pesus runs hither and thither brutishly; she gives no peace at all to the upper and lower limbs of Light. She seizes and binds the Light in the six great bodies, in earth, water and fire, plants and animals. She fashions it in many forms; she molds it into many figures; she fetters it in a prison so that it may not ascend to the height," as it says in one of the Manichaean scriptures. It is a perfect depiction of the creative process, and could be describing a pot maker, a sculptor, or a photographer. But we must also remember that this is precisely what happens when we experience something creative: we read, see, listen, and taste—and search for patterns in what we are perceiving. To read is also to create.

"It's a way of controlling the world. Creativity is really about reorganizing and simplifying. We are faced with a mystery and try to put it in order. In the natural sciences these patterns can appear as very simple mathematical formulas, like $E = mc^2$. But we all like testing, arranging, and understanding. A baby does nothing but that. It can lie for hours looking at the world, trying things out, without any external reward whatsoever. We just like it, us humans—understanding and creating order," says Laeng.

In that sense, it can also be argued that we are born creative. The creative brain takes what it finds among the chaos of patterns and colors, of quarks and nuclei, of flower petals and tree crowns—of all the dead magpies and goat horns comprising reality—and arranges it all in new patterns.

It's something most artists, scientists, inventors, and children do. Let's call it the "Askeladden" method, after the boy in the Norwegian folktale of the same name, which goes roughly like this: Askeladden goes to see a princess, who is impossible to impress. On his way to the royal residence, he picks up one useless thing after another—a dead magpie, a felling wedge, a pair of goat horns, things most people would call junk—while his exasperated brothers walk ahead, hoping to win the princess's favor. But Askeladden's seemingly irrational behavior pays off when he creates a story using the items he has collected and, as a result, charms the slightly foulmouthed and bloodthirsty princess (who has already said that those who attempt, and fail, to leave her speechless will have their ears burned with a branding iron). Askeladden makes the princess laugh so much she actually does become speechless. All his kicking around in the dirt, peering at the strange things he finds, has brought him something unexpected and new. What others see as junk becomes gold for Askeladden, and the princess is so excited that she marries him.

Creating something new involves using everything you know and can do in new ways—to construct a story or melody that makes sense, to fill a canvas with some kind of plan.

In the essay "Modern Fiction," Virginia Woolf writes:

> In any case it is a mistake to stand outside examining "methods." Any method is right, every method is right, that expresses what we wish to express, if we are writers; that brings us closer to the novelist's intention if we are readers . . . There is no limit to the horizon . . . nothing—no "method," no experiment, even of the wildest—is forbidden, but only falsity and pretense. "The proper stuff of fiction" does not exist; everything is the proper stuff of fiction, every feeling, every

thought; every quality of brain and spirit is drawn upon; no perception comes amiss.

Had Virginia Woolf heard about Askeladden, she would have seen him as a kindred spirit; someone who believes anything is possible, who gathers everything they can find and uses it for all it's worth. There's no right and wrong when you're doing something creative; all's fair in love and war, and science and literature.

I've kicked a lot of dirt around with my daughter, usually on our way home from kindergarten. We've had golden moments where a totally ordinary stone has become a treasure—something to examine and keep in her pocket—or a feather has become a key to an imaginary world. Right now, she has a huge collection of chestnuts, all sorted according to size, spread across her bedside table, drying slowly. Order comes from chaos. Things that were once useless are now important to her; what was ugly is now beautiful. It's a pattern that anyone currently in a systematic creative process will recognize. The difference is that these people will have worked purposefully for years to find interesting problems and points of attack, while my daughter has simply discovered the world and thinks that chestnuts are a great scientific find, and that feathers are unique artworks that should be exhibited on the coffee table.

"That's what it's like to create something new. Ultimately, you're looking for that sweet spot: not frustratingly difficult, not too simple—a little resistance, but not too much. The art we like exists at the frontier between challenging and understandable," says Bruno Laeng.

Those who create art or literature or music or buildings, and those who experience them, are looking for and recognizing patterns. Based on how much we know about what it is we are seeing or hearing or reading, we find the pattern either far too simple—banal—or far too complicated, and therefore irritating, frustrating. Somewhere between "Baby Shark" and a twelve-tone composition by Arnold Schoenberg, you should be able to find a type of music that hits just the right spot in your brain, something *a bit difficult* and *a bit easy*. Somewhere between *Alice's Adventures in Wonderland* and James Joyce's experimental masterpiece *Ulysses*, you will probably find your favorite piece of literature, something that will become vivid imagery in your mind. This is also one of the mysteries about the things we make—yes, about what I am writing right now: How can I make it resonate with you, as a reader? Your favorite literature is full of color, music, and smells. How can pitch-black words on dry, white paper become succulent and real? How can they give off smells and sounds and moods?

It's quite possible that an artist is a form of *synesthete*. When cross-connection of the senses occurs in the brain, it is called synesthesia—a word derived from Greek that loosely means "joined perception." A synesthete will "see" music as colors and patterns, or they will "taste" words, or, in my case, time will appear as a brightly colored landscape. When I crashed into the bridge, my synesthesia stopped. I hadn't actually been aware of it until it was gone, but it was synesthesia that had allowed me to plan for the future. My form of synesthesia is actually very common: I can "see" the coming year before me like a landscape, the weeks stretching out like a patchwork of fields in Tuscany—Wednesday-dales and Friday-peaks. And to me, numbers aren't just numbers, but shapes and colors. When my synesthesia disappeared, so did my sense of time and my ability to structure my life. But I'm not the only person to perceive space and time this way.

"Synesthesia gives my perception of time a tangible structure," says the artist Lucy Cordes Engelman. "I think that's why my work is so interdisciplinary: I'm constantly associative. All things contaminate each other like that; they are highly affective. I've studied theater and experimental film, and I write, and

I'm always combining art forms. I've basically always been interdisciplinary," she says.

"I first heard about synesthesia in 2007, when I was eighteen or nineteen and attending university. There was a vote for treasurer of a committee I was on and another student raised his hand and said that he should be in charge of the money and accounting because he saw numbers in color. I laughed out loud while looking around at everyone else, and said, 'Doesn't everyone see numbers in color?' But they all just looked at me dumbfounded, so I told my friends that, to me, the number eight was green and so on, and they were pretty surprised—and it made me feel so different and like I was living in a separate world in my mind," she explains.

Engelman, who lives in the Netherlands and the US, perceives abstract things like numbers and schedules visually—as colors and shapes—and for most of her early life thought that everyone else did too. Then came the realization that not only did this synesthesia make her very different, but few people were able to understand it. When she met fellow artist Daniel Mullen, it changed her life. Mullen's striking geometric paintings with their structured forms and vivid colors characterized Engelman's sense of time and space perfectly. Seeing them was a turning point for her.

"So we then worked together on a project to find a visual language that could share my synesthesia—and it turned into a whole series of paintings. Now I can finally visually convey my perception of time with other people, and it's amazing to open up people's ideas of what's conceptually possible in our minds. Also, meeting others who have it... I'm not so alone anymore." she says.

But the paintings didn't just help other people understand Engelman. The collaborative process brought her and Mullen so close they fell in love. Today they are married.

"Synesthesia will always be a part of me, but it certainly doesn't define me. It's just one part of my life as an artist," she says.

One of the most famous synesthetes is Solomon Shereshevsky, a man who literally remembered everything. He lived in Russia and died in 1958, and his synesthesia was extreme. Shereshevsky told the renowned Soviet scientist Alexander Luria that he once went to buy ice cream at an ice cream parlor, but when the ice cream vendor opened her mouth, thick smoke, smelling like asphalt, billowed out. Shereshevsky found her voice quite distasteful, literally, so he turned away in disgust and dropped his ice cream. He also described a particular musical note as being the color of oxidized silver, which became a steel color and then faded like twilight with a glint of silver. One single note! His synesthesia made most things almost *too* memorable, and he had great difficulty forgetting anything once he'd learned it. After trying his hand as a journalist, Shereshevsky eventually became a memory artist. Every night he would perform his "tricks," asking the audience to call out long strings of numbers that he would then remember, in one complete sequence, because they were so full of sensory impressions. The problem was he couldn't forget them again. They were virtually unforgettable.

"It's possible that we are all synesthetes when we are babies, but we unlearn this skill as we grow up. Some scientists think we remove it, that it vanishes in the slipstream when our brains undergo 'synaptic pruning' during adolescence, and the connections become fewer and stronger," says Bruno Laeng, one of Norway's leading experts on synesthesia.

Naturally, synesthesia was compatible with the myth of the artist established in the nineteenth century—the lonely artistic genius, drunk and melancholic on the fringes of society, so tormented by their otherworldly insight they were verging on madness.

"Many artists have been attributed with synesthesia; it fits the concept of the artist as a visionary, one who sees a bigger world and is endowed with unusually heightened senses. The poet Arthur Rimbaud was said to have had it, and the artist Wassily Kandinsky—who is known for his paintings that describe music," says Laeng.

Rimbaud's poem about vowels could be what started people thinking about synesthesia, perhaps more than any other thing; the letters he mentions are full of colors, smells, and tastes. There's the taste of blood and dog biscuits, the sound of buzzing flies and crashing waves:

> Black A, white E, red I, green U, blue O: vowels,
> Some day I'll speak your hidden births.
> A, black corset threaded with glinting flies
> That buzz around cruel stenches,
>
> Shadowed gulfs. E, the innocence of steam and tents,
> Spears of proud glaciers, white kings, shivering snowdrops.
> I, blushes, spitted blood, pretty lips laughing
> In rage or drunken penitence.
>
> U, the seasons, divine vibrations of viridescent seas,
> The peace of pastures strewn with animals, of the creases
> That alchemy imprints on studious brows.
>
> O, the strange and strident clarion call,
> Silences crossed by worlds and angels:
> O the Omega, violet light of his eyes!

Rimbaud later denied having synesthesia, saying that he'd lied about it, and at the age of twenty-one he stopped writing poetry altogether. After that, following a violent love affair with the older poet Paul Verlaine, he lived an itinerant and much-fabled life as an adventurer and arms dealer. But even if Rimbaud was not a synesthete, the belief in a connection between synesthesia and art prevailed.

"The ability to connect your senses like that is a good quality for an artist to have. It makes it easier to create both images and linguistic metaphors," says Laeng.

The abstract and colorful canvases of the Russian modernist painter Kandinsky are equipped with titles like *Composition* and *Contrasting Sounds*, as if they were music. Still, and this is an important point, even though Shereshevsky was demonstrably a synesthete, with perhaps the most extreme form of synesthesia known to science, his life was not devoted to creativity. If the estimates are correct—that between 1 and 10 percent of people have synesthesia, many of them without really knowing it—then there must be a fair number of synesthetes who do not become artists or work within a creative field—so there is nothing automatic about it. A new study carried out in 2016 revealed that test subjects with synesthesia scored higher when it came to openness and their ability to create mental pictures. Both are qualities linked to creativity, but not quite as much as the researchers had thought, compared with the non-synesthete control subjects.

As I said, my synesthesia vanished when I smashed into the bridge, but my head still bubbled with creativity. I was actually quite overwhelmed with ideas, so it's clearly not a crucial factor, and most artists do not have synesthesia—as far as we know.

What we are certain of is that creative people are able to see something that others cannot, a pattern nobody has noticed before. To understand a little more about how a creative person works, I contact a fellow writer, Ivo de Figueiredo. It's easy for me to talk about creativity within literature, since that's what I'm most familiar with, but what my colleague describes is equally relevant to anyone working creatively, since creativity is, at its core, about creating patterns. Ivo has, for several years, led the publisher Aschehoug's school for nonfiction writers; he has held courses and lectures on writing and is the person who has inspired me the most in my own writing career. He has written a book about the spiritualist medium Ingeborg Køber, another about his own family's origins in Goa, and a highly authoritative biography of Henrik Ibsen. Right now, he is working on a major new biography about the Norwegian artist Edvard Munch.

Before I meet Ivo, I make a point of watching *The Matrix*. He's a huge fan of this cult classic from 1999, and for one particular reason—during the film, Keanu Reeves's character, a hacker called Thomas A. Anderson, is lured in to understand something highly important. He is told to follow a white rabbit, just as Alice does in *Alice's Adventures in Wonderland*. The rabbit, in this case, appears as a tattoo on the shoulder of a young girl who rings his doorbell. Anderson follows the girl until he meets the character Morpheus (named after the ancient Greek god of dreams), who gives him the choice between two pills, a red one and a blue one. Like Alice, who can choose between drinking from a bottle and eating a cake, Anderson faces a similar choice: to take either the red pill, which will bring him greater knowledge, or the blue pill, which will return him to his old life. He chooses greater knowledge and is transformed into the superhero Neo, who can see through the matrix in the computer program controlling the world. He also develops a very strange dress sense.

"I know, the leather trench coat is a bit iffy," says Ivo. "But there's nothing wrong with Matrix-vision, because Matrix-vision allows you to see through everything, as Neo does in the film. In the end, Neo sees through, and in between, all the codes in the computer program controlling the world, and it allows him to fly and to see through walls. As a writer, you need to be equally able to see through your own material—but since your book isn't made up of calculations, and it's not nearly as logically organized as a computer program, it's more difficult to see through the structure of the text," he says.

The way I understand him, Ivo sees writing—or working creatively—as similar to building a house that you constantly renovate and extend; the living room gets turned into a bedroom and the bedroom is quickly made into a bathroom. Nothing is sacred; nothing has to be perfect. It is a constantly changing pattern. It's like knitting a sweater and then unpicking it all.

"To write is to rewrite. When you're writing on a PC, your fingers work in a similar way to what's happening in your head; you can adapt and move and edit and switch in a heartbeat. It's all about providing yourself with the myriad of words and whims flying around in the cosmos of your brain. When you write the first word, then the second, and then a sentence, you're eliminating countless other possibilities. The trick is to keep the idea and the text constantly flowing, so that you can repeatedly go back and reap the benefits of this cosmos," says Ivo, who is also working on a book about writing, based on his own experience and his course.

So a text is a universe, consisting of words and symbols that can be repeated, twisted, and changed. Once you've chosen the basic pillars for your text, building, music piece, comic strip, or film, you can move them around and give them new meanings until you have a "Matrix moment"—a kind of "aha" moment.

"Suddenly everything can fall into place and feel right. Which, of course, is a result of having done it many times before. With practice, you gain this ability to see through the structures, to move them around and see connections. I would have struggled to be a full-time writer had it not been for these few joyful moments. It's an incredible feeling. It's what also gives me confidence that when I write something stiff and boring, I can just leave it; I can allow it to stand, then go back, twist it around, and turn it into something better," says the historian.

The Munch biography Ivo is currently working on consists of short chapters, all roughly four pages long, each representing their own little text universe with a distinct rhythm and logic. He will move things and tidy things, delete and reorganize within each chapter, before zooming out and changing the whole structure of the book, reorganizing the chapters—then he'll go back in and reorganize the chapters again. Everything's constantly moving and in motion, like building several cairns and moving them between mountaintops. Okay, I'm perhaps tripping over metaphors now—writing a book is all about cairns and home renovations, and the cosmos—but you get the picture. Basically, it's not like he sits there, waiting for inspiration to come, and then throws himself, as the lightning bolt strikes, upon paper and pen to write it all down and send it straight to print. There are countless rewrites.

"It is important to establish a structure that is stable enough to provide direction, but not so firm that it becomes a straitjacket for the text. When I was working on the Ibsen biography, my idea was to write the book based on the principles 'up—down—out—in.'

"First Ibsen went *up*, then *down*, then he went *out* into the world, and then he returned to Norway and turned *in* on himself. However, this technique only really worked on paper, because it quickly turned into a straitjacket. Getting everything to fit became a real hassle; it was like I was enslaved by the idea, instead of the idea working for me. So in the end, I opted for a looser structure, which gave me more freedom and allowed the text to flow more organically, which was to structure the biography around the cities Ibsen lived in. It wasn't quite as original, perhaps, but it was appropriate, because it gave me greater freedom on so many more levels than simply the division of chapters," he says.

My own idea for organizing the book in your hand, around characters and episodes in *Alice's Adventures in Wonderland*, struck me at the very beginning. Perhaps because I had just read it, and because it is one of the most amazing and fabulous books I've ever read: pure creativity. I believed that by linking the different characters from Wonderland with the different parts of the creative process I could make everything clearer, both for me and for you, the reader. But have I not just made a straitjacket of my own? Am I overestimating the meaning of the symbols in Carroll's book? Perhaps it's not as exciting for you reading as it is for me writing about them? Then again, chasing white rabbits and mad hatters and crazy queens is quite fun.

Ivo de Figueiredo reassures me, "That's what being a writer is like; you're always looking for the buildup to a story. I constantly make stories based on my own experiences, or I'll save puns and look forward to using them."

Like Figueiredo, I also save things. Half-jokes and goat horns—it's the Askeladden method—although that doesn't mean everything is worth saving. Writing, or doing anything creative, is equally about editing, smoothing out, cutting, creating wholeness, and structure. But what are we really creating, and why do we create fictitious solutions to fictitious problems? What do we expect to do with it?

Professor Karin Kukkonen leads a large collaborative project involving literary scholars and psychologists at the University of Oslo. More and more universities are choosing to create such projects where neuroscience and cultural studies meet, because deep down, what concerns both brain researchers and literary scholars is how we understand and read and think—how we perceive reality, and how and why we try to describe it. This really applies to all those creating new descriptions of reality: architects, politicians, philosophers, artists. These stories determine how we see the world. For example, Panos Athanasopoulos and Emanuel Bylund Spångberg, two psycholinguistics professors at Lancaster and Stockholm Universities, have revealed just how deeply our understanding of time is rooted in our language. Through a series of experiments, they found that the way your mother tongue *describes* time—as long, or large, or moving from right to left or from top to bottom—determines how you *think* about time. So people who are bilingual will jump between different understandings of time.

In the same way, narratives, which move in complex ways through time and space, determine how we perceive the world—or more correctly *worlds*, in plural, since we are constantly surrounded by numerous stories and descriptions of reality.

Professor Kukkonen does not believe that there are five stories that all other stories can be built on. Nor does she think there are seven, or fourteen, or twenty. Or just one, as Joseph Campbell claimed in his 1949 book *The Hero With a Thousand Faces*, where he wrote that all myths are really the same story—a theory given some credibility when George Lucas used the book as the basis for his billion-dollar blockbuster *Star Wars*.

"I think each story is its own probability design. The action configures itself—it is its own system," says Kukkonen.

A story attempts to say something about what *could* happen, what is possible, *what if*. And a story is a building, a building all to itself, with its own rhythm, harmonies, and system of words. So why do we talk about what *might* happen, but never what *will* happen?

It is possible that these stories just leapt out of our memories. When we developed language, it was probably as a means of planning the future collectively, based on what we already knew. And they weren't just any old memories either—at first, memory wasn't there to be used on wistful nostalgia; it was to remind us of danger. Language is a direct result of our need to collaborate and swap experiences from the past in order to plan for the future, to share a vision. Initially, these visions were about tools and weapons that could ensure survival; later, they became stories about who we might be in the distant future—space travelers and cosmonauts.

"If, for some reason, it was important to humans to preserve an exact copy of the past, then that is what our memory perhaps would give us. But why would we need a replica of the past? It's the future that matters the most. The future holds potential partners and perils," says Thomas Suddendorf, an evolutionary psychologist at the University of Queensland in Australia, who in 1997 introduced these ideas in the article "Mental Time Travel and the Evolution of the Human Mind," which concerned how memory is the building block for future thinking. As further evidence for his theory, it was found that children around the age of four develop the ability to see both forward and backward in time, at exactly the same time; they are both part of the same system.

But stories about the future soon became far more than mere tales of survival. Twenty-seven hundred years ago, under a Greek starry sky, a *rhapsodist*—a man who could tell stories from memory—would sit by the fire reciting long, rhythmic poems about heroes and gods to his audience, filling the heads of his listeners with dreams of adventures beyond the horizon.

Stories anchor us in relation to the rest of the world. They give us a sense of belonging. They are also descriptions of possibilities—what we can become in the future, or in another parallel universe. Stories and creative visions are all about making the world bigger, more understandable, and organizing it in new ways.

"When you read, you encounter yourself and experience how you would react to things that never happened. Literature gives you a powerful emotional experience that you wouldn't have had otherwise. There's something very important about reading; there's something at stake, and it anchors us and creates stability. It's a probability design with its own logic, and we wouldn't read if it weren't extremely important to us," says Karin Kukkonen.

A poem by the Polish Nobel Prize winner Wisława Szymborska perfectly expresses how creativity is a possibility machine. Artists and philosophers and writers describe in numerous ways what *could have happened* and what *could have been*:

Could Have

It could have happened.
It had to happen.
It happened earlier. Later.
Nearer. Further off.
It happened, but not to you.

You were saved because you were the first.
You were saved because you were the last.
Alone. With others.
On the right. The left.
Because it was raining. Because of the shade.
Because the day was sunny.

You were in luck—there was a forest.
You were in luck—there were no trees.
You were in luck—a rake, a hook, a beam, a brake, a jamb, a
turn, a quarter inch, an instant.
You were in luck—just then a straw went floating by.

As a result, because, although, despite.
What would have happened if a hand, a foot,
within an inch, a hairsbreadth from
an unfortunate coincidence.

So you're here? Still dizzy from another dodge, close shave,
reprieve?
One hole in the net and you slipped through?
I couldn't be more shocked or speechless.
Listen,
how your heart pounds inside me.

This need to understand how the world could have been, or how our lives could have been different, has sent people in boats across the ocean, and in rockets to the moon. All stories and visions have the potential to change our lives. The words "what if" have enormous power.

Had I not read books, I would not be who I am today. Books made my world bigger. I remember sleeping in bed as a child, surrounded not by dolls and teddy bears but by books. To me, books have always been like friends, talking to me with their own unique voices. Each one with their story about how the world and life can be, they were doorways to other realities. I could have been someone else, somewhere else, in a different time. To this day, my bed is surrounded by books, and I still wake up occasionally with a triangular mark on my face, left by the corner of a book I just couldn't put down.

"There's a reason why novels have survived into the modern age, and that's because they are complex and have many different functions. It's an exchange between the reader and the text, where something you cannot predict happens. Reading contains an element of risk—and exposing yourself to something you're not used to has a value of its own," says Karin Kukkonen.

So something can happen to us when we read and write. When we read, we trigger such powerful empathic reactions that when a protagonist walks through a room or a forest, the scene will be reflected in our minds. The motor centers responsible for our movements similarly light up.

Research on mirror neurons, first observed in primates thirty years ago, has since extended all the way into the world of literary science. These neurons fire when we think about doing something, and when we see others doing that thing. We yawn when others yawn; we feel pain when we see others in pain on a movie screen.

Literature professor Gerhard Lauer believes that the same thing happens when we read about and identify with the characters in a book: we connect with them emotionally using mirror neurons. But why we do this is still a mystery. Some researchers believe that it enables us to predict behavior and cope better with the highly complex social structures of human culture. Art teaches us how life might be and how we might respond. It opens up inner worlds that are usually off-limits to us because of strict social networks and hierarchies, and it allows us to see possibilities where we previously saw closed doors.

It's a friendly gesture ultimately—to make a complex pattern out of colors and shapes, notes, or words, and then give it to someone else.

"Ideas are shared among us all the time. We live in extremely complex cultures. There are about seven thousand extant languages. Think about it! Our relations are the most complex and elaborate of all the social mammals on earth. Ideas are contagious. No one is ever alone with an idea. Nobel Prizes in science are always a joint enterprise. Could a novelist write novels without reading?" asks the author Siri Hustvedt.

Hustvedt has published seven novels, three collections of essays, a nonfiction book, and a collection of poetry, and has been translated into more than thirty languages. *The Blazing World* was nominated for the Booker Prize and won the *Los Angeles Times* Book Prize. She received the European Essay Prize for *The Delusions of Certainty*, an American Academy of Arts and Letters Award for Literature, and the Princess of Asturias Award for Literature in Spain. She is interested in how the creative process arises between the person making something and the person receiving it—it is a mixture, an in-between. The thinkers Hustvedt is interested in—Sigmund Freud, Charlotte Buhler, D. W. Winnicott, Mary Ainsworth, and Maurice Merleau-Ponty—all talk variously about "what's in between"; nevertheless, this is an uninterpreted and underresearched area. Like a placenta connecting mother and child, this "in between" is often overlooked. It wasn't until recently that researchers began examining how community works, how we bond with each other, and how we share.

"Writing books can be seen as a form of megalomania, I suppose. It is offering a gift to someone who hasn't asked for it and may not want it. It is a way of sharing my reality with someone else in the hope that between us, something new might be created. It is also an acknowledgment of our mutual dependence, and our lives together, in language. I give you a book and all I can do is hope you want it. Sometimes it works, and that brings me great pleasure," says Hustvedt.

When he received the Nobel Prize in Literature, Kazuo Ishiguro wrote:

> One person writing in a quiet room, trying to connect with another person, reading in another quiet—or maybe not so quiet—room. Stories can entertain, sometimes teach or argue a point. But for me the essential thing is that they communicate feelings. That they appeal to what we share as human beings across our borders and divides. There are large glamorous industries around stories; the book industry, the movie industry, the television industry, the theater industry. But in the end, stories are about one person saying to another: This is the way it feels to me. Can you understand what I'm saying? Does it also feel this way to you?

To make something is to reach out to other people, and maybe change their lives.

On page 240 of my own novel *Encyclopedia of Love and Longing*, I asked the reader to place their hands on the side of the book so that our hands—the reader's and the author's—could meet. It was a desperate plea from a desperate narrator to get through to the reader. "Can you feel it? Can you feel my hand pressing against yours?" I wrote. Afterward, I received many emails from people who really *had* put their hand on page 240, just as the voice in the novel had requested. I felt very moved and happy to know that it's possible to make contact with each other, across time and space and dry, white paper.

Right now, the field of *bibliotherapy* is expanding. The treatment, which uses literature to assist people with anxiety and depression, involves readers talking about books that have changed their lives, or put their experiences in a new light. Books can help people overcome trauma and heartache, anguish and loneliness. Some books have changed people's lives forever. Bibliotherapy is very often used for depression, obsessive-compulsive disorder, and eating disorders, despite there being no proof of it having any long-term effect. But if it doesn't have any long-term effect, should that matter as long as it has a short-term effect? Why stop reading? Why should only one book be enough? We need lots and lots of stories!

We don't exactly need research to prove that we can be moved by books. Reading can move someone so deeply they will place their hand on the page. My readers weren't actually touching my hand, of course, but paper—and it wasn't actually my voice they heard begging them to do so either, but another voice, one reminiscent of mine.

Literary professor Wayne Booth talks about "the implied author" and "the implied reader" and about how books are a kind of friend. Booth highlights how people in the nineteenth century referred to books as a travel partner, or a companion. It's an old-fashioned metaphor. But he believes that when reader and writer meet in a book, we enter into a contract and a relationship. Since a book is a system of characters, and a fairly limited device, in fact, an author will not be able to share everything he or she has ever thought about between the covers, no matter how much they may want to. It becomes a selection, an avatar, a set of values conveyed via black letters on a white page—"the implied author." Just think of an artist, a musician, an author, a dancer, or an actor you have been moved by: Doesn't it feel like he or she is speaking directly to you? You know who they are, right? Even if you've never met them!

"The voice in a book is not the author's own voice, of course; it is something that exists within the book's universe. A clear voice is a rare thing; very few of the manuscripts I'm sent really have one," says Kjetil Strømme Jørve, my editor at the publishing house Tiden when I write fiction.

This "voice" is what he is always looking for. An original worldview or way of looking at things, a perspective, attitude, tone, or ring. Something that resonates with others. It is then a voice.

"A voice can first appear on page 17 of a manuscript, and I'll then know there might be something there for me to work with as an editor. Then I'll read the whole thing. The voice is as much about the subject as it is about the language, I think. Technique isn't enough; it's not something you can measure by counting adjectives, or looking at the number of words in a sentence—something technical, that you can learn. It's the interplay between what's important to the author and the language the author chooses," says Strømme Jørve.

The choices made create an "implied author," a set of values, a voice, a tone the reader feels the book gives off. And a voice doesn't need to be sympathetic or pleasant for you to want to keep reading.

"The voice in a novel can be whiny, furious, cold, unkind—almost like we want to see the same existential themes handled in many different ways. We're all lonely and we're all going to die, and occasionally we might be lucky enough to experience a little love. To put it bluntly, there's not much more to be said than that. But even that can be said in so many different ways," he says.

We all need to find our voice, our angle, our perspective, in order to make something.

Comedian Laurie Kilmartin found her voice after having a child at the age of forty-one.

"I'd been a professional comedian for eighteen years when I became a mom, and suddenly I experienced a lot of situations I'd never thought I would. Like carpooling," she says.

Kilmartin says she felt like some kind of alien in the role of a mother, overwhelmed and anxiety-ridden—and that she suddenly found herself living in the most unexpected of places: the suburbs.

"Being a parent and having to relate to all these rules and expectations leaves me constantly asking myself—*why?*" she says.

For comedians, that specific question is pure gold. Having to deal with social rules that made no sense to her led to countless jokes about having kids: "I just did magic mushrooms for the first time, not to get high but to lose custody of my son," "When he was eight we had to talk about masturbation etiquette, eight! I said, 'Don't knock on mom's door when she's masturbating; that's my time,'" "I'm pro-choice, and it turns out I made the wrong choice," and "I think the worst thing about parenting is having to spend time with your children" are some of her jokes about bringing up her son alone. The voice Kilmartin found was that of a shockingly honest single parent who doesn't quite accept the rules of motherhood.

"I start with the feeling of being overwhelmed, especially my doubts about raising my son properly. So what's true in everything I write is this *feeling*, even if I alter the details. When I take my son to a comedy club, he thinks it's funny to be recognized as the son I talk about in my shows. But he's nothing like the boy I portray onstage, and he knows that I never reveal anything important about his life," she says.

Since my daughter appears in this book, it's quite possible she has played a role in shaping my voice too—you may have pictured her and the relationship we have, and it may have given you an impression of what I'm like. If you're now making assumptions about me, it's because the words in this book have given you an overall picture of the values I have, the kind of person I am. But of course, at the end of the day you know very little about me, or who I really am.

"Oh, but *I* know who you are; I *know* you!" says Beate Reed Alfheim.

Beate is one of my readers—my favorite reader actually—who, without meeting me, felt that she knew something about me based on how I write. But eventually we did meet, when she approached my table in a restaurant to tell me that she'd read, and loved, my debut novel. It made me very happy.

"I understood your strange and clever way of thinking," she said. "It's exactly how I go around fantasizing to myself. So coming over to speak to you now doesn't feel at all weird."

This happened, coincidentally, on a day when I was feeling pretty gloomy, after getting my fourth rejection for a grant to write a new novel. I was actually sitting there thinking that I should quit writing altogether, so the fact that she came over was quite significant. In fact she changed my life a little—because she made me think that I should continue writing. So, in a way, we became friends then and there, without knowing one another at all.

This is something that happens to many of us who make things and share them with the world. And it's actually more than an exchange of patterns and possibilities. Beate had read my book with *empathy*. Participating in creative activities is an empathic practice. Being part of a community; cultivating friendships, relationships, and stories; playing music for each other; creating visions as a group—it's all part of the parasympathetic response. No one sits around a campfire listening to stories if they think they're about to be eaten by a lion—when they're in sympathetic mode, stress mode, or "save yourself from danger" mode. Nor will anyone sit down and tell stories if their house is on fire. Telling a story and listening to one are surplus phenomena, parasympathetic activities that make our lives better; they increase our social competence and expand our inner world—they are nice things that we share.

"When the vagus nerve works optimally, we are healthier in body and mind, more generous and compassionate," writes Siw Aduvill in her book *Rest*.

The vagus nerve, which triggers the parasympathetic system, makes us relax and participate in the communities we so desperately need as human beings. This means that the empathy linked to any type of creative practice forms a special type of community—anyone working creatively is on the lookout for *their* community, their flock. All those who entertain want to come together with *their kind.*

"I placed the highest priority on the sort of life that lets me focus on writing, not associating with all the people around me. I felt that the indispensable relationship I should build in my life was not with a specific person, but with an unspecified number of readers," writes Haruki Murakami about the time he stopped running a bar and invested everything in writing. That's how he found his flock, a group of friends he rarely, perhaps never, meets in real life.

Creative experiences bring us together. We enjoy being at cinemas, concert halls, museums, and theaters; it creates a sense of community and prevents depression. A study that spanned fifteen years and examined two thousand people showed that when cultural experiences are shared as a group, they can have such a powerful effect that they can prevent depression. Stories, pictures, and music are a part of every culture on earth because they are so important to us; they bind us together.

Concerts are a typical place where we can experience something together, and music has an extra-strong effect on us. Making music and sharing music are powerful creative experiences that engage multiple areas of the brain simultaneously. If you think music makes your brain tingle, that's not strange, because rhythm, melody, and lyrics engage us in several areas of the cerebral cortex, and both the memory center and the limbic center (both of which are connected to our emotions) are involved. Most of the world's seven billion people listen or contribute to the creation of music. To make music is to create community.

"I can just imagine what it's going to be like. And then it happens: the light, the people standing and sitting around me. It's a very abstract moment, where I know that everything's right. It's almost like explaining a dream to someone," says the jazz musician Ola Kvernberg, who had performed at Rockefeller Music Hall in Oslo the night before we met.

Kvernberg has received multiple awards, including a Norwegian Grammy. He started playing classical violin as a child, and as a fourteen-year-old ended up in third place in an Italian violin-playing competition. Then he discovered jazz. At the age of seventeen, he showed up at the Django Festival in Oslo, 230 miles from home, and asked if he could take part in the program. The festival manager was so excited after seeing him play, he immediately brought Kvernberg onstage for the opening performance with the renowned jazz band Hot Club de Norvège. Soon after that, he appeared on his first record with the Dutch superguitarist Jimmy Rosenberg. Since then, he has released nine albums.

With each album release, there's a tour—as there is now—where Kvernberg meets his fans and plays at packed venues where something magical happens: he enters the legendary "flow" state. It's something many musicians talk about—a sense of losing yourself in a perfect musical moment; an encounter between the audience and musician that doesn't otherwise happen; a moment of intense presence.

"I don't think I can actually remember the moments where I'm in the flow. I only remember the moment that I 'wake up' and I'm no longer in it," he says. It's like a dream, because you don't have full control, but at the same time, of course, you do.

The very latest brain research shows that experienced jazz musicians are actually in a kind of dream. The test subjects of one study were experienced jazz pianists who had automated the motoric nature of playing music to such a degree that they could jump into what is called the DMN, the so-called "default mode network"—daydream mode—while playing. What made these musicians special was that while they were in daydream mode, they were also using executive function, which provides direction, concentration, and planning.

Daydreaming isn't enough if you are going to make something; you have to make something that has artistic form, that has a voice and a direction.

"If I'm asked what the next most important quality is for a novelist, that's easy too, focus—the ability to concentrate all your limited talents on whatever's critical at the moment," writes Haruki Murakami.

"Without that you can't accomplish anything of value, while, if you can focus effectively, you'll be able to compensate for an erratic talent or even a shortage of it," he continues. "Everybody uses their mind when they think. But a writer puts on an outfit called narrative and thinks with his entire being," he writes, about the effort and concentration required to find direction and focus.

Direction and form are things that can take years to find. Finding the voice and the story is not always easy; there's no standard way of doing it, no particular trick.

"Some writers work very slowly. I've been working with one author for eight years, and we still haven't reached the end of the book. I get sent one chapter a year," says Kjetil Strømme Jørve.

Well, I won't be spending eight years writing *this* book. I need a deadline! Otherwise, the project will carry on in much the same way it started—as an eternal pretext for tidying up the attic. That's right. Writing a book can be a great opportunity to tidy up your loft. In the time I've set aside for writing this book, I've instead completely redecorated the office (quite essential for writing, of course—don't argue!). I've made no end of mouthwatering dinners that required days of preparation. I've also arranged an evening of 3D-photography, participated in organizations, read 438 news articles about things I'm not terribly interested in, polished my nails in five different colors, and become strangely attracted to jogging—despite the fact that I hate running just as intensely as Haruki Murakami loves it. Perhaps I was subconsciously thinking that if I started doing what Murakami does, I would become as good a writer as him—or perhaps I was trying to exhaust my inner critic, or simply to run away from the book I was writing. I have also tidied my wardrobe and, most recently, as I mentioned, started investigating the possibility of cleaning the attic. It was then I really understood what was going on: I was procrastinating.

Had I not known any better, I might have panicked. But I no longer fear procrastination. I now know that it can be a creative strategy. According to the philosopher John Perry, so-called "structured procrastination" can be productive. In fact, Perry wrote an entire book about it, *The Art of Procrastination*, to avoid having to do all the other tasks lying on his desk. He believes that it's quite possible to be productive even when you are deferring the actual hassle of writing something. He analyzes his procrastinating and one of the key points he finds is that he tries to avoid doing the important things. And why does he not want to do what's at the top of his list? Because he is a perfectionist—and to perform a task that will result in something that's less than perfect will always seem impossible, until he is forced to complete the same task when he has run out of time. And when he is out of time, there will be no chance of doing it perfectly, but he *will* at least manage to deliver something, albeit late. In the meantime, he will have done numerous other things, further down the list, that are less important but important enough, such as writing a book about procrastination, for example. In that way, procrastination can work quite creatively and enable you to sneak around the Queen of Hearts without her noticing. What's important is to fill your to-do lists with lots of points, because if the list has only one thing on it, then a skilled and determined procrastinator will avoid doing that task by simply doing *nothing*, paralyzed by their own perfectionism, and the Queen of Hearts will then get a firm stranglehold. If everything is dependent on just one big task, the whole house of cards will collapse, according to Perry.

"Can't we avoid the emotional turmoil and the waste of everyone's time that these perfectionist fantasies lead to?" he writes, exasperated with himself.

He recommends that we lessen our need for perfection by prioritizing, breaking things down into manageable tasks, and listening to some nice music at work. And of course—stop worrying about producing something perfect. Just produce!

So where do I begin if I want to build a house, compose a piece of music, create a work of art, or write a TV series—or a book? There has to be somewhere to put the first letter, the first brushstroke, the first note—because when you're faced with the Great White Unknown, only intrepid adventurers like Nansen and Shackleton press on fearlessly. How am I supposed to get started?

"You can deliver on Thursday," says my editor for this book, Erik Møller Solheim. Solheim is also the editor for the comedian Else Kåss Furuseth and her book about going to the psychologist, which she spent six years writing. "He's very kind," Else confides, and I think, Yes, he is actually. Kindness is important when you are writing a book. Had I been stressed and afraid about talking to my editor, this book wouldn't have happened. Had he thrown me into a sympathetic response that weakened my immune system and left me unable to sleep at night, I would never have been able to do this. I need a friend—a fairly strict friend—when I write.

"Setting a deadline is an art in itself, one I still haven't fully mastered—whether I'm doing it for other people or for myself," says Erik Møller Solheim.

Some writers routinely miss deadlines, have notorious problems delivering on time, procrastinate, or just sneak off. Other writers thrive on tight deadlines.

"It's the thing I hear most often from my authors," says editor Kjetil. "'Can you set a deadline for me?'"

"It's a way of restricting yourself," says editor Erik. "It frees you from the responsibility of writing a masterpiece. So if it doesn't turn out to be an amazing book, it's no longer your fault. Any restriction on creativity is a benefit to you. When everything is possible, nothing becomes possible, and there's a risk of you imploding," he says—then adds, "So you'll deliver the manuscript on Thursday then?"

Thursday seems totally impossible. I sigh heavily while thinking about my attic. Then I think about how I don't want to implode. I think about my muse, Kristopher Schau.

I decide to start writing, and just pretend it's improv theater. There's no time to write anything perfect; my inner critic should just give up! Although she'll demand that I find the perfect opening sentence, of course. And if I'm unable to find that, I'll be unable to write anything. "So you may as well give up," shouts the Queen of Hearts furiously. It's then that I find this scene from *The Plague* by Albert Camus, describing Grand, a writer, who has come up with the first sentence for his book: "One fine morning in the month of May an elegant young horsewoman might have been seen riding a handsome sorrel mare along the flowery avenues of the Bois de Boulogne." But he isn't happy with it; it seems he never will be, either:

> That's only a rough draft. Once I've succeeded in rendering perfectly the picture in my mind's eye, once my words have the exact tempo of this ride—the horse is trotting, one-two-three, one-two-three, see what I mean?—the rest will come more easily and, what's even more important, the illusion

will be such that from the very first words it will be possible to say: "Hats off!"

I know plenty of writers who get stuck on the first sentence. So I make a decision: that it's okay if people don't shout "bravo" at the first line. And *that* allows me to get started; I'll be able to rewrite and tinker with it whenever I want until the editor puts his foot down—it'll be fine—and I write, "So I hit the wall, literally." I'm sure I will have to change it later, but it doesn't matter. The thing is, this whole book about creativity is one long procrastination project, an attempt to sneak off and do something other than what I'd planned to do. Obviously, what I *should* be doing is writing a novel!

4 | Wonderland

OR: THE MYSTERIOUS SOURCE OF CREATIVITY, DMN.

"Would you tell me, please,
which way I ought to go from here?"
"That depends a good deal on where
you want to get to," said the Cat.
"I don't much care where," said Alice.
"Then it doesn't much matter which
way you go," said the Cat.

IT'S NO COINCIDENCE that this is chapter 4. We're halfway through the book, and I'm now beginning to understand what it's really about—it's about getting lost. It is here, in the middle of a dark and misty forest, on a slowly vanishing path where there's no GPS signal, that I find them: my daydreams.

Highly creative people don't actually do what you might expect or what mythology suggests: they rarely use drugs, they're not always drunk at some bohemian café, or acting weird in public. Very many of them work hard, for prolonged periods, and are very focused. But they don't work like factory workers or administrative staff, and the methods used by the creatives I interviewed for this book are quite special. They will shower, or sit in the bathtub, go walk in the forest, or hang about town; they'll go cycling or skiing; they'll ride trains or fly on planes, or meditate their way to being creative. They fumble about; they'll have no direction at the start, and sometimes they'll stay directionless all the way to the finish line. Some lie awake at night, and some use being half-asleep as a technique. None of these methods seem like an actual technique either, because there doesn't seem to be one magic word or trick I can steal for my own use. I now understand that I need to chase several white rabbits into dark rabbit holes.

And there, in the dark, I begin by taking a nap. Daydreamers have to be good at dreaming, and I've found that many people use their after-dinner nap as a creative tool. For example, in an article for the *New Statesman* in 1963, the author Doris Lessing writes:

> When I sleep after lunch, there is nothing unplanned about it. First I must feel the inner disturbance or alertness that is due to overstimulation, or being a little sick or very tired. Then I darken my room, shut all the doors so the telephone won't wake me (though its distant ringing can be a welcome dream-progenitor) and I get into bed carefully, preserving the mood. It is these sleeps which help me with my work, telling me what to write or where I've gone wrong. And they save me from the fever of restlessness that comes from seeing too many people. I always drift off to sleep in the afternoons with the interest due to a long journey into the unknown, and the sleep is thin and extraordinary and takes me to regions hard to describe.

Lessing's article describes "a room of one's own," referring to the book by Virginia Woolf. It is a room where you can breathe deeply and write freely—a room that, in the past, was extremely difficult to find if you were a woman who had children and household duties taking up all your time. For Lessing, that room was her bedroom.

"I spend a lot of time in the bedroom. Bed is the best place for reading, thinking, or doing nothing. It is my room; it is where I feel I live," she writes in the article. In 2001, Lessing received the Nobel Prize in Literature. Perhaps because of, rather than despite, her little naps.

Haruki Murakami also swears by his afternoon naps. "Usually I get sleepy right after lunch, plop down on the sofa, and doze off. Thirty minutes later I come wide awake," he says.

But how can resting be a path to success? How can *being bored* or *getting lost* help?

"That's something my father taught me—that I'll get bored and that I should work with people who are better than me," says comedian and author Else Kåss Furuseth.

"I was often bored as a child. And when I complained, my father just said, 'Good.' Now, when I'm working, I'll constantly fiddle around, right up to the day of the premiere. It'll be a complete mess and chaos, with occasional chinks of happiness—like when you have the bouquet of flowers in your hands because you've made it, the show is done, and the premiere was a success. But there's never any guarantee it'll end up like that, so that's of course not why I do this. But I'm not really sure if I can properly describe what's going on, because it all feels very unstructured."

Else Kåss Furuseth's performance at Norway's National Theater had people laughing at subjects like depression and suicide, and her book, which is loaded with comic reflections on human vulnerability, has long been on Norway's bestseller lists.

She clearly has no problem talking about mental illness, as many of us do. But she is embarrassed by how she works.

"I mess around so much! The only thing that works for me is the deadline. A show is a show and I have to be at the TV studio; it has to be recorded, otherwise it won't happen. So I just have to be there! And I can't actually do anything; I can't sing or dance! But if I just have to be there, and then I'm there, I do it anyway."

Many creative people, in addition to resting and fiddling around, spend a lot of time being bored. It's a recurring theme when I start researching—being alone, directionlessness, and boredom.

"I get all my best ideas on trains or at train stations—and airports, back when I used planes more often," says my editor Erik Møller Solheim, who is also an author.

"'Being bored is good for you,' my dad would always say. Boredom and quiet time were things I got for free from my mom and dad. After dinner, there would always be an hour of silence. They would rest. And this would, of course, happen in the middle of the living room, which meant that I had to be quiet, play with my Lego, and listen quietly to the little cassette player in the playroom. An hour of monastic peace, as I remember it. It's something I'm eternally grateful for," says Ivar Johansen, a musician, hit songwriter, and TV presenter who goes by the name of Ravi.

There's clearly a link between creativity and boredom. We live in a culture where boredom and lack of stimuli are seen as signs of something being wrong. But if you are stimulated, entertained, or constantly in other people's company, it's difficult to have your own thoughts. Boredom is unpopular; very few people go deliberately looking for it. Yet it appears to be something we need, perhaps more than we think. After hearing so many established creative people talking about boredom and directionlessness, I realize it's a path I need to follow.

"I wonder what it does to us, the fact that we're never forced to be bored anymore. You only need to put your hand in your pocket and you're connected socially to Facebook. So we're never confronted by boredom," says Lars Svendsen. To demonstrate, he reaches for his smartphone, which two seconds later is in his hand.

Lars Svendsen, a philosophy professor at the University of Bergen, made his debut as a writer with the book *A Philosophy of Boredom*. It was first published in Norwegian in 1999, then in English in 2005, long before smartphones hooked us permanently to the internet, an event we could call the mass extinction of boredom.

The book struck a chord; since publication it has inspired both researchers and philosophers, several of whom have continued working in the field of boredom. Svendsen actually wrote the book after becoming so incredibly bored—while doing his PhD on the philosopher Immanuel Kant—that he wanted to find a new way of pursuing a career in philosophy. Since then, he has written thirteen books, which have been translated into twenty-seven languages; there are 105 translations of his books in total. So, in this case, writing about boredom paradoxically opened the door to an interesting life as a writer.

"There's something quite useful about boredom that makes it good for creativity—having a fairly boring life full of habits and routines allows you to focus on things other than organizing your life. Look at Ibsen or Immanuel Kant, whose lives were both totally full of routine," Svendsen points out.

But there's something else about boredom that makes it fertile ground for creativity. To experience boredom properly during the interview, the philosophy professor and I sit and wait at Oslo's Central Station. Now *that's* boring! A huge railway station like this is full of massive amounts of boredom. Thousands of people have paced back and forth here, waiting for trains that are delayed, and ones that never came; it's as though you can breathe in every single sigh, every single groan of boredom that's ever been expelled; the air is thick with monotony and frustration.

The PA system interrupts our conversation at regular intervals with announcements about delays for those traveling. It's a constant reminder of how incredibly boring it is to wait for a train.

Dr. Seuss writes about the "Waiting Place" in the children's book *Oh, the Places You'll Go!*, which, among other things, is about being bored and stuck:

> Waiting for a train to go
> or a bus to come, or a plane to go
> or the mail to come, or the rain to go
> or the phone to ring, or the snow to snow
> or the waiting around for a Yes or No
> or waiting for their hair to grow.

What is it about boredom, these gaps in time where we are lost and unable to focus on anything but our own inner monologue? The moments where the only thing we encounter is our inner self?

"When you're sitting at an airport or a railway station and hear that your flight or train is canceled, you're thrown into a meaningless existence. Your journey is the only reason you're spending time in that place, and when you are no longer doing *that* anymore, you lose direction," says Svendsen.

He thinks this is a good thing. Being bored means risking something, something you cannot see when you are constantly being entertained. Social media, shopping, porn, reality TV shows, and iPad games keep us so occupied we don't notice what is happening in our own inner world.

> Vainly I have sought an anchorage, not just in the depths of knowledge, but in the bottomless sea of pleasure. I have felt the well-nigh irresistible power with which one pleasure holds out its hand to another; I have felt that inauthentic kind of enthusiasm which it is capable of producing. I have also felt the tedium, the laceration, which ensues. I have tasted the fruits of the tree of knowledge and relished them time and again. But this joy was only in the moment of cognition and left no deeper mark upon me.

This is how the philosopher Søren Kierkegaard described aimless boredom in 1835. Kierkegaard did not like boredom. It worried him.

"When bored, you are continually thrown back in on yourself. So for Kierkegaard, boredom was the root of all evil. And for the Fathers of the Church, it was the worst of all sins—boredom was the source of all the other deadly sins, because it allowed human weakness to come marching in unimpeded," Svendsen explains.

So no, boredom, or in Latin, *acedia*—much like curiosity, *curiositas* (which I wrote about earlier)—was not appreciated in Christian theology either. It was a condition that allowed the sins to gain a foothold and a firm grip of the heart. Boredom would be followed by *laziness* and *salaciousness* and everything else that was forbidden during the Middle Ages. I'm starting to wonder if the seven deadly sins, perhaps, might be the path to creativity after all.

Boredom is therefore associated with losing direction, a kind of void that can be both important and uncomfortable. I actually spent a lot of time feeling bored at Lars Svendsen's house, when I looked after his two cats many years ago. They too were clearly bored, rising lethargically from the sofa as I let myself in; they would then sit in the window gazing out at nothing. The two cats, Lasse and Geir, resembled the ever-expectant Vladimir and Estragon in Samuel Beckett's play *Waiting for Godot* from 1953. The play brilliantly illustrates aimless waiting and meaninglessness, since, although Vladimir and Estragon expect him to arrive at any moment, Godot never appears. And that's how it was for the cats. The cat sitter (me) was of no importance, just something that roused them from their sleep and into a state of boredom. I could almost hear them talking:

Vladimir (Lasse): Say, I am happy.
Estragon (Geir): I am happy.
Vladimir (Lasse): So am I.
Estragon (Geir): So am I.
Vladimir (Lasse): We are happy.
Estragon (Geir): We are happy. (*Silence.*) What do we do now,

now that we are happy?
Vladimir (Lasse): Wait for Godot (Lars Svendsen).

"I've said in the past that animals cannot be bored. But they can, of course. Animals care about food, and about catching their food—it gives their lives meaning and direction. Captive animals should be given tasks related to obtaining food," says Svendsen, who has just completed an essay about animal boredom and has written a book about understanding animals.

Svendsen now has a dog, who never gets bored—and who finds its meaning in life through cuddles and going for walks. But it's possible that humans benefit from boredom in a completely different way than animals. We humans need boredom, and have made more and more allowances for it, yet we still don't seem to appreciate it as much as we should, even though many thinkers since Kierkegaard have considered it important.

We are now living in a kind of paradox of boredom: there is a great duality in contemporary Western life. Since the sixteenth century, the discoveries and inventions that have made life better—the amazing technological innovations we have made—have paradoxically made life worse. We now have machines that do things we once spent days trying to do, such as obtaining food. We now buy our food instead of making it. We drive instead of walking. Our houses are heated by electricity instead of wood we've chopped up and carried home ourselves. It is a friction-free, uneventful life: What does this do to our brains, which are actually built for problem-solving, finding new and weird solutions, having ideas, making associations, and learning? What's clear is that instead of using our free time constructively, or by doing nothing at all, we actually spend it entertaining ourselves. For example, we spend an average of three hours a day watching TV, and 470 minutes a day on the internet. We don't prioritize *getting bored*. Boredom is something we avoid, if we can.

An experiment in the United States conducted on fifty-five people was published in the renowned journal *Science*. The test subjects had to sit alone in a room, in complete silence, without doing anything. They were told not to cheat, but they were allowed to bring their smartphones into the room with them. And almost everyone cheated. They were then deprived of their phones, and suddenly the fifteen minutes in the room became very uncomfortable for them. But it was the final experiment that was the most telling: this time the test subjects were provided with a way of giving themselves a small electric shock while they were in the room, and they were allowed to test this on themselves first. Before entering the room, fifty-two of the fifty-five subjects said they found the shock very uncomfortable and would pay good money to avoid being shocked.

Once inside the room, the subjects were allowed to use the electric shock as a distraction during the fifteen minutes of silence. And astonishingly, many of them did precisely that—a third of the men and a quarter of the women—several of whom had just stated with full conviction that they would go out of their way to avoid the electric shock. One of them shocked himself 190 times.

"This says a lot about how difficult we find being in our own company," says Lars Svendsen resignedly, and slightly amused.

"Boredom involves your mind being full of things you don't care about, or lacking things you do care about. It's not tragic, or painful, sad, or dramatic—it simply boils down to not knowing what to do, because there's nothing you actually want," says the philosopher.

Naturally, not knowing where you are going can lead you to completely new places. Directionless boredom contains something deeply creative, but something uncomfortable too.

According to the author Fernando Pessoa, boredom is "to suffer without suffering, to want without desire, to think without reason." Both Wittgenstein and Heidegger saw boredom as the source of philosophy itself, a place without direction and desire, from which thoughts about the world could grow.

When you are feeling bored, when you are thrown back at yourself, your entire inner world is magnified. And by today's standards, your inner world is not all evil and sinful; there's a lot in there that's fun and weird, existential and nice, and that can be used in a creative process. When you are bored, you need to find something new to focus your attention on. Boredom is a doorway to an incredible world.

"Boredom is the gateway to mind-wandering, which helps our brains create those new connections that can solve anything from planning dinner to a breakthrough in combating global warming," writes Manoush Zomorodi in the book *Bored and Brilliant*, which is about the advantages of being bored.

After boredom comes daydreaming.

In 1862, a young girl was dozing on a riverbank in the shade of a tree, while her sister read aloud to her. In the story, the girl Alice vanishes down a rabbit hole into a magical world, perhaps the same world that Doris Lessing visited as she lay on her bed in London around a hundred years later. Alice's Wonderland is a dreamworld that's, perhaps, very important for creativity. While sleeping is crucial to us being at all functional—and dreaming is a mystery scientists have yet to solve—*daydreaming* has for several years been called "creative mode," a kind of light version of dreaming.

Just to be clear: waking dreams, the DMN (the resting network I described in the introduction), mind-wandering mode, and daydreaming mode all describe the same thing. This network in the brain was discovered by the brain researcher Marcus Raichle. He and his team were able to examine the brain with a so-called fMRI, which he also helped to develop. In fact, he scanned people while they were not stimulated and lying calmly with their eyes closed—primarily to get a "baseline," a measuring point he could use as a comparison when examining the same people doing special tasks.

fMRI, in combination with cognitive psychology, has revolutionized brain research in recent years. With the help of these two tools, researchers can "see" and measure people thinking—the device allows them to see the large clouds of activity that form where the brain's network of neurons has flooded with oxygen, as people work out specific tasks and participate in testing. Because that's how the brain is—nerve cells are not individually active; they cooperate, to give us ideas and memories, to solve problems, and to understand. We now talk increasingly about networks, rather than centers. Yes, we have a brain center for vision and hearing, but to understand how we think, it is more important to examine the networks and interactions in the brain. Executive function is one such network. The default mode network (DMN) is another.

"What's been getting a lot of attention lately, due to its important role in creative thinking, is the DMN," says Professor Joy Bhattacharya.

"The DMN means that we, often independently of external stimuli and tasks, can wander in our minds. But we also have something very special that happens during artistic activity; the executive function cooperates with the DMN—so that you get a thought migration with direction. Usually, the DMN and the executive are mutually exclusive," the professor explains.

So you either daydream and look in on yourself, or direct yourself outward, focused and concentrated. Jazz musicians, however, do both of these things simultaneously when improvising. As mentioned earlier, jazz musicians will also play their instruments automatically, without concentrating on the motoric nature of playing, and can therefore enter what is commonly known as "the flow," where concentration and daydream mode are elevated to a higher level. It's like a guided tour, where instead of wandering around the forest with no goal or purpose, you have a path to follow, a focus, a direction—a white rabbit to chase through Wonderland—while directing your thoughts inward, associating and daydreaming freely.

This connection between executive function and the DMN applies to anyone working creatively, of course—not only jazz musicians. When I'm writing, for example, I go into my own world, but I never forget that I want my story to go somewhere. Writing with no direction rarely leads anywhere good. Following every whim and association is perhaps closer to what occurs in the DMN, but it doesn't necessarily produce great art, because your material will probably lack both form and direction. In the DMN, the flow of thought is not targeted.

"Much of our work consisted of checking if what we'd seen was perhaps due to an error in how we'd examined the brain," Raichle explained afterward, after the groundbreaking discovery of the DMN.

Eventually, his research team understood that what they had found was a separate network, not something they had registered by accident. They then began investigating what this network actually does. Raichle and his team's discovery of the DMN has sparked many new research projects over the last ten years, and in that time, their original article has been cited over six thousand times. Countless neurologists around the world are currently mapping what the DMN really consists of. One thing that initially baffled them was how difficult it was to pinpoint exactly what it does for us.

"Paradoxically, it's more active when you're *not* engaged in a task that has a purpose," Raichle explained. His team found that when we're not focused on solving problems, understanding the outside world, sorting impressions, or being in conversation—things that require executive function to be in a high gear—the brain automatically switches to the DMN. It is therefore called the "default" mode network. These two systems work like our muscles, like biceps and triceps: when one is tensed, the other has to relax, and vice versa. When the executive function is resting, the DMN is active, and vice versa. Hans Berger had observed something similar in the 1930s when measuring electrical activity, using electrodes applied to the skulls of people who were either resting with their eyes shut or solving a particular task: the brain is in alpha during wakeful rest periods where you wander aimlessly through your mind, and in beta when you are in executive mode. Having your brain in alpha is important for "aha" moments—something I experienced in chapter 2, of course. There is probably a connection between alpha waves, the DMN, relaxation, and ideas.

Understandably, researchers have not been terribly interested in the brain's daydream network. The world of medicine and neuroscience has examined the entire human body in terms of its function and usefulness, and *purposelessness,* you would assume, has no obvious function. It would seem strange to examine it at all. However, we spend about 50 to 60 percent of our waking time in the DMN according to researchers in the field, and we use nearly the same amount of energy as when we are thinking purposefully—and this uses up a large amount of the body's resources. Our brain is a calorie sponge: 20 percent of the calories we eat and 20 percent of our blood is used by the brain, despite the brain accounting for only 2 percent of our body weight. So this purposelessness must be very useful to us; otherwise *Homo sapiens* would never have developed an organ requiring so much sustenance. When the brain became what it did, every single calorie required was associated with hard work.

Part of the DMN is "concerned with social behavior, mood control, and motivational drive, all of which are important components of an individual's personality," wrote Raichle in 2015 in a bigger article looking back on his surprising discovery. The DMN is closely associated with the hippocampus, which preserves memory, and in the DMN, memories are processed and inserted into our life history. It also turns out that the DMN is important for consolidating memories, which is crucial for retaining them.

It is natural to assume, given all that we know about "aha" moments and alpha waves, that the DMN is also particularly important for having good ideas. And rightly so: new research at the University of California, Santa Barbara, shows that important "aha" experiences occur in the DMN. Researchers asked both writers and other researchers to report their "aha" experiences, a fifth of which were said to have occurred while daydreaming. The "aha" moments that occurred in the DMN were seen as far more important than the task-driven "aha" moments, and showed that there is a difference between a large and a small "aha" moment: one is the "Einstein's theory of relativity" moment, which appears like a daydream after a "storm in the brain," while perhaps the most common is the "Vincent van Gogh" moment, which is the little "aha" you get while painting a picture of a bench. When we're under pressure or solving problems, we will have "aha" moments, of course, but the big, groundbreaking ideas will most likely come when we're not looking for them. The place to go for fruitful and sometimes uncontrolled brainstorming, and for truly groundbreaking thoughts, is the DMN.

If it is true that we have about four thousand thoughts a day (I've read estimates of up to seventy thousand, but it's obviously quite difficult and slightly pointless to measure *numbers* of thoughts), it's quite possible that up to half of them are spontaneously created in DMN mode, which means that we very often assemble and test our ideas and thoughts while in the DMN, with no plan or direction whatsoever. It's somewhere we just associate and dream. But it will sometimes reward us with totally new and unique ideas—and "aha" moments—like when my daughter jubilantly exclaimed, "Chocolate-milk-juice!" or when Trude Lorentzen suddenly realized that the empty bedrooms she was looking for were those belonging to the children who died at Utøya.

But none of this is vague speculation that we *might* get more ideas from being in the DMN; researchers have observed an *immediate* effect—in an experiment conducted using Guilford's Alternative Uses Task test, the participants did far better after taking a twelve-minute break where they daydreamed than when they first had to complete an equally long and demanding task that engaged their memory. In a study from 2017, researchers found a connection between having good cognitive skills and spending a lot of time in the DMN. One of the reasons for this, they believe, is that people who think fast and solve tasks quickly will go into the DMN when they are finished, and will also remain in this state longer. Schoolchildren who solve a problem, for example, and then patiently wait for the other children to finish, will gladly drop into the DMN because there is nothing else to do. It may look like they're inattentive and not concentrating, but in reality they are doing an important job: they are processing information, consolidating memories, understanding themselves and others, and they just might be having some good ideas.

A quick glance at the history of humankind shows that in the beginning we weren't entirely free to wander aimlessly in a perpetual daydream; our species needed to collect a lot of food and be constantly on guard for external threats. In more recent centuries, however, the opportunities for many of us have only gotten better, and what was once a luxury, or done secretly while collecting raspberries and hunting, is now something we can do unhindered and for hours at a time. And the results of this are the visible and tangible changes in human culture: from the nineteenth and throughout the twentieth century, those of us privileged enough in the Western world have become more secure, while humanity has simultaneously brought about an explosion of innovation. We have successfully fought a number of lethal diseases; many of us now live in safe, well-functioning societies that protect us so much that we rarely, if ever, need to be in *sympathicus* (not that it seems to have prevented us from being stressed). We have built a fortress around human life in the form of a growing number of inventions and discoveries. And this volume of creative ideas has increased, thanks to our *nondirection*. So nondirection (with creativity) has been one of the secrets behind humanity's successes, and our dominance as a species. It is paradoxical and strange that we can thank nondirection for so many things, and at the same time have so little appreciation for it. Without daydreaming, there would be no moon rockets or knitting patterns, no televisions or Velcro, no *Mona Lisa* or theory of relativity, no Pippi Longstocking and no Beethoven's Fifth.

Beth Lapides, comedian and founder of UnCabaret, knows what laying the groundwork for good ideas requires: being without direction. That's one of the reasons she does yoga every day. UnCabaret came about as a nonxenophobic, nonmisogynistic alternative to late-1990s stand-up. Her idea for it—arguably her best and most important one—came suddenly, and it changed her life and modern comedy forever.

"That's how I've seen it for the last few decades, at least. I've always relished telling the story about this instantaneous decision. Because light-bulb moments are exciting and illuminating. I actually believe I get a kind of contact high from remembering this epiphanic blink," says Lapides in her audiobook *So You Need to Decide*. However, she began to investigate further and discovered that before this great "aha" moment, there were lots of minor decisions and events that had paved the way to UnCabaret.

"I realized that a sequence of decisions had gotten me to the point where I could blink, and that there should be a corollary book to *Blink*, called *Stare*," she says, referring to Malcolm Gladwell's book *Blink*, about the sudden moments of insight that can permanently change your life.

Lapides therefore prefers talking about the opposite of blinks, what she calls "the stare." While "blinks" are impressive and grand, "the stare" is more comparable to lying in wait for your prey before catching it.

"The thing about staring is that it's not as flashy as blinking," she says.

"But it's the long slow burn of gazing and grazing, *the stare*, that makes these sparkly instantaneous decisions possible. Seeing the stare was in itself a blink moment," she writes.

While searching through her life story for the roots of her idea for UnCabaret, she discovered that there were many of them—from the boring and painful months she spent being treated in hospital for a rare blood disease when she was a child, to her experience of the misogynistic stand-up comedy scene. This is the groundwork of good ideas: waiting, thinking, having no direction.

"It might also have all started when I got the role of the Cheshire Cat, the first role I played onstage. Perhaps that's what planted the seed of what would become UnCabaret?" she says.

Her first stage appearance in a children's production of *Alice in Wonderland* was never meant to revolutionize comedy. But that's what's great about being nondirected: you never know where you might end up.

So let's take a closer look at what it means to be nondirected.

Tracing my own mind-journeys along the overgrown pathways of my brain, I revisit a few old advertising jingles, some quotes I want to use in this book, and then panic as I remember I need to do my tax return. I then take a stroll to an embarrassing moment I had recently, before suddenly remembering that I need to buy flowers for the backyard, and then come across some nice memories of swimming in the fjord. I think a bit more about the DMN, and an article I want to read about creativity and botanical gardens, then try to formulate some concrete sentences for the book you have in your hand, before picturing someone, most likely you, reading the book, and hopefully having your life changed by it.

All this happens in the few minutes it takes me to walk to the bus.

Then a memory from a holiday I once had in Thailand pops into my head, and I'm baffled by how the beach and palm trees can just appear, out of the blue, so vividly that I can hear the sea and smell the jungle.

It dawns on me that, if cultivating my creativity is the goal, perhaps I need to do more to facilitate being undisturbed in the DMN. Greater thinkers than me have done this: "I procrastinated for so long that I would henceforth be at fault, were I to waste the time that remains for carrying out the project by brooding over it. Accordingly, I have today suitably freed my mind of all cares, secured for myself a period of leisurely tranquility, and am withdrawing into solitude," wrote the philosopher René Descartes in 1641, revolutionizing philosophy and science with his work *Meditations on First Philosophy*.

But finding a quiet space to retreat to is more difficult in the 2020s than it was in the 1600s. So I do something I never imagined myself doing: I pay to lie in a water tank for an hour of so-called "floating." In the experiment I described earlier, the test subjects had enough trouble staying quiet and bored for fifteen minutes. Now, I would have to stay totally, totally quiet for sixty minutes. Four times as long. It's a bit of a hippie thing to do, but it's also very much in line with what I know about creativity: that it thrives when the brain is in daydream mode, undisturbed by external stimuli. Maybe I'll experience something new and exciting? I won't be able to go anywhere—I'll be naked and trapped—so I'll be completely alone with my thoughts and associations.

My friend Guro, who is currently working on a PhD in psychology, takes me to the studio she regular visits. "I love it!" she says (she does this quite often). "Nobody can reach me in there, so it's very relaxing." It's somewhere you can literally float inside your very own think tank.

To be fair, there's something else about flotation tanks that makes people pay to spend time in them: they contain salt water.

It's remarkable how many great ideas have occurred either in or near water, so I try not to be too skeptical. I think about Lewis Carroll sailing through Oxford in a boat on the river Thames, as his story about Alice and the White Rabbit comes to him. I think about the Mole, in the children's book *The Wind in the Willows* by Kenneth Grahame, who starts his adventure by wandering aimlessly and then ends up down by the river, which he has never seen before:

> All was a-shake and a-shiver—glints and gleams and sparkles, rustle and swirl, chatter and bubble. The Mole was bewitched, entranced, fascinated. By the side of the river he trotted as one trots, when very small, by the side of a man who holds one spellbound by exciting stories; and when tired at last, he sat on the bank, while the river still chattered on to him, a babbling procession of the best stories in the world, sent from the heart of the earth to be told at last to the insatiable sea.

And so begins one of the world's greatest stories about a mole and a water vole and a toad. It is water that sets things in motion. So too with Archimedes, Trude Lorentzen, Richard Feynman, and the composer Gustav Mahler—who got the idea for one of the movements of his acclaimed symphonies while rowing a rowboat.

According to Mathew White, who is working on a research program called BlueHealth at the University of Exeter, water has a particularly calming and health-promoting effect, even greater than the effect parks and forests have on us. His research group has looked at the effects of water—both seas and fountains—on human mental and physical health in eighteen countries. One experiment involved asking twenty thousand smartphone users to report if they were happy and, if so, precisely where they were at the time—and it became clear the most commonly given locations were on the coast. White thinks that being close to the sea is a particularly good way to prevent rumination, brooding, and negative thinking, all of which are associated with depression.

One characteristic of depression is that it impairs our ability to picture the future and recall detailed memories from the past, and naturally this harms our imagination and our ability to depict things. Depression typically makes writers less productive, according to research. So I think spending time in or near water will be great for my creativity. It will make me parasympathetic and happy, free to explore my own mind and to literally drown my inner critics and demons. I'll become more aware of the world and have the energy to pursue my dreams and visions, which in itself is an antidote to depression. To read, write, make art, and explore scientific theories is to observe the world's many details and colors—which in turn connects us to life. It is the opposite of depression, which makes everything look gray and dull and cloaked in a dark fog. Yes, I know all this perhaps sounds just a little bit New Agey, but I have science on my side. Happiness lives in the details. Happiness lives in the waves—and the poet and artist Miek Zwamborn agrees.

"Five years ago we left the Netherlands in order to be closer to nature, and now live on a Scottish island on the west coast with less than three thousand inhabitants. Before then, my strong and intense relationship with nature had been detached from the intellectual realm of the art world—it wasn't something I used in my work because it seemed somehow forbidden during my studies. It was as though I had to split myself in two," she says.

One of Zwamborn's books, *Seaweed: An Enchanting Miscellany*, presents an international cultural history of seaweed, from its earliest use as medicine and transportable food for hunter-gatherers; to animal fodder along coastlines; inspiration in painting, fashion design, film, photography, and poetry; and as a product to farm and cook with.

"A lot of my work has been inspired by the sea and forces of nature; there's so much I want to try to understand. Looking after this small corner of the island and studying it feels like a huge privilege and responsibility too. It's a fragile environment and I am very grateful that we now run the KNOCKvologan residency, involving other artists, writers, and composers connecting to this wonderful place," says Zwamborn.

Through his research with BlueHealth, Mathew White also found that parks and forests ranked very highly on the happiness scale; they make us feel relaxed and are therefore a good way of entering the DMN. Even before this research became known, the late neurologist and author Oliver Sacks had a strong feeling that there were two things that were good for our mental health: music and gardens. Sacks believed them to be highly beneficial for our health, and the two things that made him most creative.

"In forty years of medical practice, I have found only two types of non-pharmaceutical 'therapy' to be vitally important for patients with chronic neurological diseases: music and gardens," he writes. When he lived, he always visited botanical gardens in the cities he traveled to. (This was, as you perhaps realize, the same article I'd thought about on my way to the bus earlier in this chapter.)

"As a writer, I find gardens essential to the creative process; as a physician, I take my patients to gardens whenever possible," he writes. "In many cases, gardens and nature are more powerful than any medication," he continues.

Research shows that people who live or spend a lot of time in the countryside are less prone to depression. At a population level, access to nature is hugely important to public health. In addition, there might not be such a big difference between walking and writing a book and improvising a piece of jazz: your mind will wander; you will grasp things around you mentally, not in a stressed or targeted way, but in an observational way with no clear direction.

Nevertheless, any journey usually moves between two points. You'll go from one place to another—without being stressed, but with direction. I think of all the times I've walked in the botanical gardens and come away feeling happier, and the time I visited Virginia Woolf's garden in Sussex, which was so overgrown and full of life. Virginia Woolf, the queen of the stream of consciousness, the woman who better than anyone could describe her train of thought—her DMN. Maybe it was due to the months she spent in her summerhouse, in her beautiful, sprawling garden, while letting her mind wander freely? Or perhaps it's actually the other way round: maybe creative processes imitate walks in the forest?

Mina Adampour is a doctor whose family comes from rural Iran, although in Norway they have never had time for walks in the forest. But when her brother became ill, he began walking in the countryside to help himself recover, and it changed Mina's life too.

"I started researching, and I then realized that there's something about nature that's especially good for us. People living on the west side of Oslo still have a longer life expectancy than those on the east side. And there's one big difference: those living in west Oslo either go walking in the forest more often or they go to their cabins—out in the countryside," she tells me, as we walk through the woods near my home.

We walk through Svartdalen, a sliver of ancient forest cut through by the river Alna, which flows through Oslo's east side, leaping and dancing past old abandoned factories, and out into the fjord. This narrow valley was once inhabited by people who worked hard and looked after their children, but probably had neither the time nor the energy to be creative. Yet hard, physical work can sometimes allow your mind to wander, so perhaps they daydreamed quite a lot?

Right now, at least Mina and I have more than enough time to walk through the woods in search of a parasympathetic mode and good DMN. The best thing about what we're doing is that it makes you healthier. At the University of Michigan, researchers found that spending just twenty to thirty minutes in nature, three times a week, is a very effective way of bringing the body's elevated levels of stress hormones down to a normal level.

"Forest walking brings us a sense of well-being, and in men, it increases the levels of adiponectin, which helps prevent cardiovascular disease. Studies have also found that walking in nature reduces stress faster than walking in urban areas," says Adampour.

Walking in nature has been linked to having more energy, which is something you don't have if you work long hours and have no time left for anything other than surviving. In such cases, it's easy to fall into a vicious cycle of bad food and sedentary leisure time in front of the TV—if you actually get any leisure time. If you feel very stressed, going out for a semi-directionless walk in the woods will seem almost impossible.

"Another experiment looked at gardening and happiness," said Adampour about what she had discovered. "Because when we dig soil, we inhale the bacterium *Mycobacterium vaccae*, which makes our brain cells secrete the mood-altering hormone serotonin, which makes us happy. Mice that were injected with this bacterium displayed less anxiety-driven behavior, and it improved their memories."

Nature has made Adampour more creative as well. She has written a collection of poems through which she has allowed nature to speak; with a looming climate crisis, she felt that nature perhaps needed someone to express its pain. As well as writing poetry, she started researching as much as she could about how the forest affects our health, and is now writing a book about it from a medical perspective.

Research on forests has gained momentum in recent years, perhaps most of all in Japan, where so-called "forest bathing," *shinrin-yoku*, has been practiced since the 1980s. There is also a lot of research in the US about how nature affects our health.

I hear the sound of a waterfall, and I'm reminded of how good ideas thrive near water. When I was at my worst, when my head really wasn't cooperating, I took my concussion for a relaxing break at a spa hotel. After two days with nothing but the sound of running water in my ears, I felt significantly better.

"Think about it. Practically all screen savers use a nature photograph. Because it calms us down. You can even buy an app that plays the sound of rivers and waterfalls. You can, of course, just go out in nature and stimulate *all* of your senses at the same time, and that will calm you down. Stress is incredibly harmful, for both the mind and body. Yet we learn almost nothing about it in medical school," says Adampour, fervently.

My walk with the doctor feels like we're practicing being in both executive function and DMN at the same time. We are in motion, but we are not stressed. We're embraced by green trees, breathing fresh air, and allowing our minds to wander as we discuss art and science. Saturday mornings are a lot better like this.

• • • •

AT THE FLOTATION studio, I'm given a towel and assisted by a young male employee. He says that the water in the tank is even saltier than the Dead Sea and advises me to smear myself with Vaseline to stop any little cuts I might have from stinging when I'm immersed. I think about the time I visited Israel and swam in the Dead Sea and how impossible it was to get my body under the water. I remember seeing an Israeli bobbing around like a cork, with a newspaper that remained totally dry on his stomach.

In the little flotation room, there's a shower for me to use before and after the session. Inside the tank itself, which resembles a little white plastic boat, I can choose to have it completely dark if I want—and I can pull the doors shut as I enter. Then I float on the surface, stark naked, in water that stays constantly at body temperature. The effect is supposedly like being a baby in a mother's womb. What do babies think about? What do fetuses dream about in the dark and the heat, surrounded by amniotic fluid, the muted sounds of the mother's body, and the world waiting outside?

As I climb into the tank, more practical thoughts begin swirling around my head. "What if a guy with a machine gun walks in now; I wouldn't be able to run away!" "I really shouldn't have eaten so many chickpeas for dinner." "It's a bit annoying how my head constantly sinks just beneath the surface so my ears get blocked. Getting water in my ears is the worst thing I can think of."

"I *have* to relax!" I tell myself repeatedly. "And I have to stop *telling* myself to relax; it's not relaxing!" I tell myself just as many times.

I feel almost unbearably self-conscious in my totally silent room. The rumble of the tram passing by now and then reinforces my feeling that the whole world is in motion—except me. What am I actually doing here? This is ridiculous. My ears are continually letting in water, so I sit up and try to shake it out. Then I lie back down in the tank and just wait for them to play the short piece of music to indicate the end of the session. The silence is absolute. I close my eyes and sink deeper into consciousness. Shapes and colors appear behind my eyelids. According to research, this can happen when you are in deep relaxation, when your brain waves are in alpha. The study, at the Queen Mary University of London, was conducted on an artist who creates art from the images experienced when meditating with her eyes closed. Researchers passed an electric current through the artist's brain and then controlled her brain waves. When her brain was set to alpha, her mental pictures became very sharp and clear, while the very intense gamma waves produced images that were blurrier, although they lasted much longer. I come to the conclusion that since the images I'm observing are sharp and momentary, it must mean that I've succeeded in calming my brain to the alpha wave level, to daydream mode. Of course, none of this is terribly scientific, given that the study was done on just one artist and without a control group—and just how scientific can I be when I'm floating naked in a water tank? Besides, who knows exactly how my brain works? I take a deep breath and allow myself to relax a little.

It's silent.

Oh, so silent. Plop, plop, plop.

My toe brushes against something. Was that the side of the tank?

Silence.

Silence.

Salt water in my ear. Both my ears.

Silence.

Ugh! My toe!

I drift off again. Then I suddenly remember everything I've done and learned recently. On any normal day, I have an endless stream of thoughts flowing through my head, ones that usually involve remembering to buy milk or to pay bills. Now they seem magnified. There I am, bobbing around in the tank, dreaming about completing this book, imagining what it will be like to receive the Nobel Prize in Literature. I lie there naked and practice my acceptance speech. (It is humble but interesting, wise and funny in equal measure; people laugh; they're all a little choked up. I see it all quite clearly in my head. Even Mette-Marit, Crown Princess of Norway, wipes a tear from her eye.) Research on the DMN has shown that I'm not the only one fantasizing about winning the Nobel Prize. Psychology professor Eric Klinger at the University of Minnesota has found that two themes appear frequently in daydreams: one is a hero fantasy, the other a martyr fantasy, both of which involve a protagonist who overcomes a series of difficulties and is eventually recognized for who he or she really is. The hero fantasy is linked to success, which for me means getting the Nobel Prize. Or maybe it's really a martyr fantasy—you know, after years of hard work and suffering, I'm finally recognized as the genius I really am.

The DMN has many functions and is perhaps crucial to how good we feel about ourselves. It is not just a creative network, but a state of mind, where we get to know ourselves and the world around us, where we consolidate memories and plan for the future.

"An important task for the default mode network is integrating memories of life events in a self-relevant way," says Professor Raichle, who is now over eighty years old and has received the prestigious Kavli Prize for his accomplishments within brain science. He believes that the DMN helps us understand ourselves, to become human.

"Our daily memories play a crucial role in helping us form a model of the world we live in, and therefore enable us to assume something about and predict the future. I suspect that the default mode network provides the integration that makes this possible," he says.

The DMN is where any thought beginning with "what if" occurs. And "what if" thoughts are the start of a creative process: a poem, a novel, a work of art, a piece of music, a scientific theory. What if there was a girl with red hair and pigtails who lived alone with a spotted horse? What if this symphony began with the sound of fate itself knocking on the door?

I love daydreaming. I like just walking around lost in my own thoughts, at the "party in my head," as a friend calls it. Kurt Vonnegut refers to something similar, in a scene in his novel *Breakfast of Champions* where the protagonist is sitting in a café: "Can you see anything in the dark, with your sunglasses on?" the waitress asks him. "The big show is inside my head," he replies stoically.

As a child, if someone ever asked what I was doing, I would often say I was "just thinking"; I was rarely bored because I always had the whole interior of my head to play in. I clearly remember slowly meandering home from school, walking through puddles, or picking up leaves, twigs, and pebbles while my head buzzed with thoughts, songs, and ideas, fairy tales I'd heard and fairy tales I'd made up.

Now my four-year-old does exactly the same thing. Getting her home from kindergarten in a straight line is impossible. She dawdles, she sings, she walks off in the wrong direction, points at things, asks about things; she has so many whims and thoughts I'm not always able to keep up with them.

"I'm allergic to sidewalks," she said one day before randomly walking into the road, which was luckily empty at the time.

"Mama, do you know how I knew it was going to rain? It's because I came down like a raindrop to you and turned into a baby, so now I know a lot about the weather," she said patronizingly on another day, explaining her meteorological insights.

She is like a live demonstration of how the DMN probably works. According to some researchers, children and young adults are in DMN far more often than adults, and it makes sense, given that they have to learn and remember far more about the world, and also try to understand far more about themselves and other people than we have to as adults, when we know ourselves and the world much better.

It also strikes me that my child is ridiculously unstressed. She doesn't care at all about bus times or bedtimes, career decisions, or her performance at kindergarten. It's like she is totally satisfied with who she is and what she does, and has no problem forgetting time and place or anything related to goals and plans, in favor of whatever weird things are going on in her head.

The DMN is strongly linked to introspection and the absence of external stimuli. This doesn't mean that external stimuli are of no importance; they can bring you into DMN as well, just as the ocean does. The truth is, as researchers in Berlin and Bergen have found, it's easier to go into DMN if you listen to a piece of sad introverted music, rather than something more outgoing and happy. The researchers conducted three different studies with as many as 216 test subjects participating in the initial trial.

First, they made a definition of "sad" and "happy" music, based on what many of the subjects thought. The results showed that far more introspection, self-reflection, and mind-wandering occurred when the music was sad, provided that all the other factors were the same—such as the tempo of the music. So when the philosopher and procrastination expert John Perry recommended putting on some nice music to work to, his advice was perhaps better and more scientifically sound than he first realized. And when Marsilio Ficino believed that the music of the lute could soothe the melancholy of the creative genius, he wasn't entirely wrong, nor was neurologist Oliver Sacks. Sad music will make you more introverted, and therefore more able to get in touch with your inner voice.

But right after my crash, listening to music was impossible. Today, I still get tinnitus, and loud noises are ear-piercing in a totally different way than before. Sudden bangs or screeching brakes can make me feel physically sick, which is fairly common among people who have suffered concussions: the thalamus, which receives signals from the senses and passes them on to the brain, can be disrupted by head injuries. Even when someone with a concussion shows no sign of injury whatsoever on an MRI image, it's still entirely possible that the connections in their brain have been destroyed and disrupted. Research into concussions has only really just started, but it has already been found to change the relationships between the brain's networks and put them off-balance. This can result in a higher risk of depression and problems with dizziness, in addition to obsessive-compulsive disorders, sleep problems, irritability, headaches, fatigue, memory loss, and concentration problems. Concussions have been found to increase activity in the DMN, which can include both positive and negative effects; increased activity in the DMN is, in some way, linked to depression and ruminating.

Researchers in Boston and Mexico have just found that meditation affects us similarly to improvising music: meditation creates a tighter connection between the DMN and executive function, which is also linked to mental health. It is logical to assume that other types of "boring" activities, such as cycling, swimming, or yoga, will put you in a similar state of directed daydreaming, where the mind wanders freely while the body does something you need to be partly concentrating on. The same thing happens when we color in drawings (which was recently quite trendy) or knit—we do something we can master so well that our minds can wander at the same time.

I read about the relaxing effects of knitting in the *New York Times*. Dr. Herbert Benson, who has written *The Relaxation Response*, believes that the repetitive motion of the knitting needles can put you into a relaxed state, quite similar to the one that occurs during yoga and meditation. A trained knitter will have lower blood pressure and secrete less of the stress hormone cortisol while the needles are moving. In Toronto, Karen Zila Hayes has established a knitting therapy program that includes Knit to Quit, for smokers, and Knit to Heal, aimed at helping people through crises such as cancer diagnoses or serious illness among close family members.

What's clear is that it is here, in DMN, that one of the most important parts of the creative process unfolds, despite it still being hard to conclude decisively what this network is—given that neither fMRI nor alpha wave readings are entirely accurate measurements of what's going on in the brain. But no matter how we measure it, we have to daydream—we cannot create something totally new unless we can handle being alone in our own heads. We might experience it as restlessness, mild anxiety-driven boredom, fiddling around, aimlessness, free-flowing associations, or an inner conversation, perhaps a discomfort or—as Einstein described his breakthrough with the theory of relativity—a storm in the brain. It's hard to account for the time you use, because it never really seems like you're doing anything useful. But now we know what's going on during all this "time wasting"—you are roaming around in your own weird head. Still, I've spoken to a lot of creative people while writing this book who are quite ashamed of this part of the process.

I, for one, am very ashamed.

The author Thure Erik Lund has long experience of being alone in his head. "The rich source of my writing is when I was young, walking around my parents' farm. I had to work a lot, and spent a lot of time by myself," he explains. In his essay collection *Romutvidelser* (Room expansions), he writes about the necessity of being alone: "If you isolate yourself from the world for a long time, in order to write about the world, you can quickly be seen as detached from reality, and worst of all outdated, reactionary, and dangerous," he writes, pointing out that literature is subject to absurd demands regarding commercial use and marketability. I sometimes feel that same pressure myself, the same need to be topical, relevant, or *useful*. I feel so ashamed when I just wander around *thinking*; it feels so useless.

So where does this shame come from?

Perhaps the ancient ideal that work should always have a purpose, reinforced by the Protestant work ethic, is what causes all the shame and anxiety in the creative people I've spoken with. Just a few generations ago, most people were deeply religious, poor, and uneducated. They saw their children, or those of their close relatives, die of illnesses we now take for granted as being curable. The only place they could turn when faced with the brutalities of life was their faith. My great-grandmother, who bore nine children and worked on a farm her entire life, couldn't exactly go to her GP for a referral to a psychologist when life got tough; she went to the priest. The culture back then was permeated by the ideal of hard work, which was perhaps necessary considering the tough circumstances. In *The Protestant Ethic and the Spirit of Capitalism*, the sociologist Max Weber describes the "iron cage" that emerged from Protestantism and laid the foundations of modern-day capitalism: a good Protestant should work hard and save money, not for their own benefit, but for reinvesting. A flourishing business would be a sign that God was with you—although there was no guarantee of that either. Life had to consist of hard work and as little fun and games as possible.

"For when asceticism was carried out of monastic cells into everyday life, and began to dominate worldly morality, it did its part in building the tremendous cosmos of the modern economic order," writes Weber.

This entwinement of work ethic and a Protestant image of God can be seen in modern Western society today, according to Horst Feldmann, a social economist at the University of Bath. Feldmann examined data from eighty countries to see if there were differences in employment rates between nations that have different religions. And yes, there were! Employment is *significantly* higher in countries where Protestant Christianity is the most widespread religion, compared with countries where other religions are dominant.

"Protestantism rather generalized the virtue of hard and diligent work among its adherents, who judged one another by conformity to this standard," says Feldmann, summarizing the relevant historical scholarship.

Daydreaming, in a culture like this, was the opposite of purposeful work and, of course, wholly unacceptable. In a Protestant world, hard work is about honoring the glory of God; it is deeply existential and meaningful. There is no time for "just thinking" or messing around.

Ibsen's fanatical priest figure, Brand, said in the play of the same name, "My God is of another mind, / A storm, where yours is but a wind, / Where yours is deaf, inexorable, / All-loving, where your God is dull." The religious fervor Brand preached didn't mean that people in strict Protestant cultures didn't daydream while they rhythmically and semi-purposefully shoveled manure or built a factory. But I do believe that this sense of shame was born from that fervor.

And it's a feeling I know well. I went around for months wondering what this book was really about. Terrible months, full of self-loathing, where I just couldn't find the direction. I felt like a bum, a slacker, looking for something that seemed both vague and unfathomable: Boredom? Mind-wandering? The vagus nerve? The parasympathetic? Gardens? Rabbits? None of these things were tangible. I just saw myself as lazy.

"Love, wisdom, grace, inspiration—how do you go about finding these things that are in some ways about extending the boundaries of the self into unknown territory, about becoming someone else?" writes Rebecca Solnit in her collection of essays *A Field Guide to Getting Lost*.

"Certainly for artists of all stripes, the unknown, the idea or the form or the tale that has not yet arrived, is what must be found. It is the job of artists to open doors and invite in prophesies, the unknown, the unfamiliar; it's where their work comes from, although its arrival signals the beginning of the long disciplined process of making it their own," she continues.

Keats called this "negative capability," describing it as being "when a man is capable of being in uncertainties, mysteries, doubts, without any irritable reaching after fact and reason."

The first time I read Solnit's book, I was reminded how, like many people, I find venturing into the unknown, without knowing what it entails or what kind of transformation it will bring, quite scary.

As Dr. Seuss puts it in *Oh, the Places You'll Go!*:

You will come to a place where the streets are not marked.
Some windows are lighted. But mostly they're darked.
A place you could sprain both your elbow and chin!
Do you dare to stay out? Do you dare to go in?

If daydreaming is a way of encountering the unknown—yes, if the unknown and fantastic is how you make a living—then it's something you have to take seriously, even if it's scary. You have to endure being there with it, amidst the silence and uncertainty.

Anna Fiske is an award-winning author and illustrator whose work has been translated into over twenty languages. She was educated at art school in Stockholm and has nearly always worked creatively—except for the time she cleaned floors and a short period when she had to copy maps, neither of which was for her.

"My inner world has always been my own space. I've always enjoyed playing, letting my thoughts drift, and wandering around in my own head. It's something I nurture. I'll often sit in the bath and have a good think," she says. I first met Fiske twenty years ago, when she had just published a very moving graphic novel about suicide. Since then, she has published a large number of books—many of which are lined up on a shelf in her studio, along with her latest children's book, *How Do You Make a Baby?*, which has been hugely successful. She has written books for adults, children, and young adults—almost an entire shelf of books, about everything from trees and insects, to puberty and anxiety, to bottoms and happiness.

"I'll often glance at all the books up there, and yet—every time I start writing a book, it feels like I'm doing it for the first time, like I've never tried it before, like anything can happen," she says. "Right now, I feel like I'm running low and need a recharge, so I'm going south, to the coast, to be by the sea. That usually helps. The sea is so alive and, at the same time, it doesn't want anything from me. I just stand there looking at it, taking it all in."

It's not just books at her studio either; there are all sorts of things: drawings, small figurines, a pile of wooden scraps that—with a little paint and some felt-tip drawings—have been transformed into colorful little houses with people inside. A giant cardboard pine needle that reaches the ceiling leans against the bookshelf, and on the floor, there's a papier-mâché pine cone the size of a small dog. There's a rabbit in her studio too—although it's pink, not white. There are plastic dogs, a doll's house, and a bulletin board covered in small tourist badges from little Swedish towns.

"I found these badges at a flea market. And then I drew my own characters onto them," she says while pointing to them. They are strange, bright-pink animals with funny faces, wandering around the pictured cities.

"Working creatively is good for my psyche. And I look forward to going to work every day," she says. "But for me, it's important that this isn't just a job; I don't want to go home from the studio and stop thinking. Creativity brings me so much joy. So I challenge myself to see all the strange things around me. I don't necessarily want it all to turn into something, like a book. It's more like my own creative gym: How can I see the world in new ways?"

We drink coffee and eat cinnamon buns. Her new project, which spent several years maturing before being put to paper, is spread all over her desk.

Fiske will normally think a project through to the very end before starting to draw it properly. She'll imagine it and knead it into shape in her head, long before sitting down at her tidy little drawing board. By then, it's been finished—in daydream mode—and is ready to be put on paper.

"It is difficult sometimes. I sometimes feel I should just work as a bus driver instead, although I'd be constantly terrified driving a bus! Besides, I just love marveling at things and vanishing into my own imagination. I often do nothing. I don't *want* to fill up all my time. I'll rest a lot, and I'll often sit on the sofa at home, or lie in bed and look straight ahead," she says.

With age, Fiske feels more open. For her, getting older means becoming more sensitive. When she is in this heightened state of sensitivity, she feels like she is taking everything in, all the joy and pain around her.

"Sometimes all the emotions can make me feel quite exhausted; it can be a lot to take in. But it's what I want. When I don't feel like that, I become quite sad, even though I know the feeling will return. It's such a powerful force for me, working creatively. It's my place," says Fiske.

Another author I know shields her inner life with a determination verging on fanaticism. Ida Jackson is a blogger, a mother, a children's book and nonfiction author, and an internet aficionado. She is an extremely systematic daydreamer.

"For me, creativity is like a mental superpower that I can use for whatever I want. I don't get writer's block and I don't fight with my inner critic much, although I know it's there. I just have to allow myself to play somehow," she says.

Jackson's creative regime involves getting up at 5:30 every morning and writing three pages. They are not meant to be shown to anyone, and they have to be written by hand—and on paper, to prevent them being edited or sent to anyone by email. These three pages are full of doubts, irritations, random thoughts, and dreams she's had during the night, and they allow her to see what her inner critic has in store for her that day. It's almost like a map of her DMN. She calls it "written meditation."

"Since creativity is no longer a secret lover I have outside of my internet job, I've tried taking it both more and less seriously. Of all the things I do—besides being with my husband and child—what's most important is allowing myself to dedicate my life entirely to being creative. There could be a zombie apocalypse tomorrow and we could all be dead soon, so for that reason I don't care what people think; although it's all just fun and nonsense, anyway. I'll approach the project in a deadly serious manner, but with the occasional fart joke thrown in," she says.

Jackson's weekly schedule also includes some sacred hours of play. It's impossible to reschedule any meeting she has booked with her inner ten-year-old. These are opportunities for her to do the things she always wanted to do as a child, but forgot to do as an adult.

"I look at my child, my four-year-old—he is creative plutonium. His world is huge and totally electric. I haven't taught him to play; of course, it's an impulse that comes just as naturally to me. So I have to make time for my own inner child and do the things I once thought were super exciting. Stickers, for example. That's fun! Taking a sauna. Sitting on a swing. Or lighting a fire. Lighting a fire is great fun—I'd forgotten how much! I wanted to return to how I was as a ten-year-old, before puberty messed everything up for me," she explains.

Finding this inner world full of joy and self-awareness, this exciting place where stories appear, is also described by Virginia Woolf in the novel *To the Lighthouse*:

> For now she need not think of anybody. She could be herself, by herself. And that was what now she often felt the need of—to think; well, not even to think. To be silent; to be alone. All the being and the doing, expansive, glittering, vocal, evaporated; and one shrunk, with a sense of solemnity, to being oneself, a wedge-shaped core of darkness, something invisible to others. Although she continued to knit, and sat upright, it was thus that she felt herself; and this self having shed its attachments was free for the strangest adventures. When life sank down for a moment, the range of experience seemed limitless.

In her nonfiction book about neuropsychology *The Shaking Woman or A History of My Nerves*, Siri Hustvedt describes something similar: "Clearly, a self is much larger than the internal narrator. Around and beneath the island of that self-conscious storyteller is a vast sea of unconsciousness, of what we don't know, will never know, or have forgotten," she writes.

Of course, strings of associations and self-reflections don't amount to an exhaustive description of the brain; they are one of many conditions.

"We order our memories and link them together, and those disparate fragments gain an owner: the 'I' of autobiography, who is no one without a 'you.' For whom do we narrate, after all? Even when alone in our heads, there is a presumed other, the second person of our speech," writes Hustvedt, describing the mysteries connected to her flow of thoughts.

I met Siri Hustvedt when she visited Oslo to talk about her new novel *Memories of the Future*, a book in which Alice from Wonderland is alluded to in the name of a dog.

"Are you sure you want to use *Alice in Wonderland*," she says, when I tell about my book. "You know it's one of the most interpreted books in English?"

Of course, Lewis Carroll's book is more than just a banal allegory of creativity. It wasn't meant to be a story about inner critics and "aha" experiences and executive function and the weird and wonderful DMN. He wrote it to entertain Alice Liddell, and any other children with an appreciation for first-rate fantasy, using characters drawn from the depths of his peculiar brain. Carroll quite deliberately passed Alice's adventure on to her older sister, to the reader in fact: "But her sister sat still just as she left her, leaning her head on her hand, watching the setting sun, and thinking of little Alice and all her wonderful Adventures, till she too began dreaming after a fashion."

In Siri Hustvedt's new novel, she gives a key to the reader: Alice's key to Wonderland. "Hold out your hand. I am giving you the keys. One story has become another," she writes, echoing the conclusion of Carroll's book, and, in doing so, reopening the rabbit hole for the reader. We are all storytellers. We are all part of creativity, and to measure human creativity and the DMN is deeply problematic, thinks Hustvedt.

"There is a tendency in the science world to take an abstract concept and harden it in physiological terms. It happens with fMRI. Movement is stopped and turned into a map. We must also remember the artificial conditions in a lab. Tell people to daydream in an fMRI machine and they may feel pressure to perform rather than let themselves go. It's also quite noisy there."

Everything I've written so far might seem alluringly simple; the brain seems almost like a mechanism, the DMN like a garden you can just enter and harvest ideas from. But the human brain doesn't necessarily function that way, and humans are continually evolving. Ultimately, much of the brain is still a mystery, an island kingdom we can only map from the coast.

Siri Hustvedt fears that we believe too much in brain science and don't allow ourselves enough skepticism.

"Scientific models hope to capture a dynamic reality, but they have mostly been static maps. New Bayesian models for the brain are founded on the idea that the brain is a predictive organ. They are complex and ingenious, but no one really knows if they mirror actual neurological processes or not. The danger arrives with confusing the model for the actual thing," she says.

Modern neuroscience, like all modern science, has its roots in the West and began in the sixteenth century. It is mathematical, based on repeated experiments, and seeks to reveal the laws of physics in order to manipulate and control the world. These laws of physics enable us to say something about the future, not just to describe the past. Natural law applies to all times. Through research, we have been able to use formulas and experiments to advance civilization. These secret laws and connections in the natural world have been transformed into steam engines, telephone lines, vaccines, and computers. But scientific methods are far more difficult to apply to our wildly associative, emotionally controlled, intricately woven, fantastic, and creative brains.

"Mechanistic models for understanding biology have come under fire. Can we think of the brain's resting state and its generation of reveries, thoughts, and ideas as reducible to distinct mechanisms?" asks Hustvedt, who has spent a lot of time acquainting herself with modern brain research.

Nevertheless, memory research can tell us quite a lot about what happens when we daydream. We imagine the future based on what we already know; we place our aspirations there and create horror scenarios based on our nightmares. We want to know what the future has in store because we are worried it might be deadly—and it is, of course, deadly; we are all going to die eventually. But despite the inevitability of death, we are frightened by it and do everything we can to control it. At the same time, the history of humankind has shown that our future is open and unpredictable, and that we can turn the most incredible fantasies into reality.

Humans once walked around the African savanna with only spears in their hands. Now we are connected to each other with John Logie Baird's seeing telephones. For thousands of years, we dreamed about flying to the moon; now we have done it. Lewis Carroll also talked fervently about self-writing ink, and now we have it! I can write half a sentence on my cell phone and it will finish the rest for me (perhaps one day I'll be able to write, "So I hit the wall. Literally," and the rest of the book will write itself). Thanks to fMRI, we can now see how oxygen-rich blood fills the brain's various networks while we're thinking. But it's still impossible to explain why the sea makes so many of us happy and calm, or what it means to have direction. It's impossible to explain what consciousness and personality really are, or where you want to go, and why. Or what it actually is you're pursuing when you chase a White Rabbit.

Siri Hustvedt spent a whole year working on a novel, and it died in her hands. A DMN with direction might sound very simple, but being creative involves following a map that doesn't exist, and looking for north or south in a country where nobody knows where they're going.

"The novel I'd been working on was stillborn, but I couldn't just give up. I had written 150 pages, and there was something important missing. I had the feeling that it was growing sideways, not moving forward, but what does that mean? It means that it lacked the drive all narratives must have, a push that feels as if it is going ahead. But what is that sense of direction in a book? To write is both to control one's work and to lose control of it. That forward motion is a subjectively felt sensation, which is rather mysterious," Hustvedt says.

Suddenly, she got an idea that saved everything.

"Ideas come suddenly, when I least expect them to. They happen when I let go," she says.

Her new idea was just as physical, funny, and weird as her first one was thoughtful and cerebral—a young woman who moves to New York for the first time hears voices on the other side of the wall in her apartment, and the adventure unfolds from there: a story about two young detectives, a woman with a dark history, and a dead child.

The moral in the genesis of Hustvedt's idea is this: if everything is under control, nothing new will happen. The creative place is where you end up while dozing on the riverbank, not when you're hunched over a laptop and trying too hard. It is presumably how we manage to find our own emotional truth and propel ourselves into the fairy tale.

Siri Hustvedt once taught a group of writing students, which made her slightly frustrated. "What no one can teach to a writer is emotional truth, the most important force in all of fiction—and this, of course, varies from writer to writer. The students in my class didn't write badly, but many of them wrote from the outside in. They thought they could impose rules on their texts, that there were specific methods that could be used to make the work good. But there are no rules in art. One can clean up ugly sentences until the end of time, but it will not give a text force, feeling, or emotional truth," she points out.

If you want to find your own, individual truth, you must tolerate being alone. You can't be constantly stimulated; you must dare to sit quietly. Because of what I've learned about daydreaming, I now put my cell phone in my pocket far more often if I'm waiting for a bus or train, or if I'm queuing at the store, or sitting on a beach. Instead, I'll try to find my dark inner core, and stay there, to explore what happens when I'm *totally silent*. I try to pursue my thoughts unsystematically; I plan novels and children's books and essays and articles; I talk to myself and listen to myself. From the outside, it might look like I'm just dawdling around, mumbling, reading books, sniffing raspberries, lost in my own thoughts. I'll walk in the woods in the middle of the day instead of sitting at the computer. But despite the shame and self-loathing, I'll continue to think and write. It's something I can't stop doing.

"What we know about creative people is that they are motivated from within, and not due to social conventions or financial reasons," says Professor Bhattacharya.

This explains why most people working in so-called "creative professions" do so despite salaries that are often way below average. The median salary for an author in Norway is about 13,000 US dollars, which may well be what I earn from this book—which I spent over a year writing. Working as a writer is a risky business, where the financial rewards are as uncertain as Captain Ahab's odds of catching Moby Dick, or Alice's hopes of beating the Queen of Hearts at croquet: chances are, you'll lose.

But my reward is getting to spend time in Lewis Carroll's amazing Wonderland, which for me has almost become a description of the DMN. It's a place where everyone is crazy—the Mad Hatter invites you over for tea parties, oysters sing on the beach, flamingos get used as croquet mallets, and a large caterpillar puffs away on a hookah. Or I could use Carroll's phrase "large bright thing" when referring to the DMN. Lewis Carroll depicts all the unpredictability of daydream mode—the state you can access only when you're *not trying too hard*—in his book *Through the Looking-Glass* (the follow-up to *Alice's Adventures in Wonderland*), where Alice has a conversation with a sheep that is knitting:

> "Things flow about so here!" she said at last in a plaintive tone, after she had spent a minute or so in vainly pursuing a large

> bright thing, that looked sometimes like a doll and sometimes like a work-box, and was always in the shelf next above the one she was looking at. "And this one is the most provoking of all—but I'll tell you what—" she added, as a sudden thought struck her, "I'll follow it up to the very top shelf of all. It'll puzzle it to go through the ceiling, I expect!"
>
> But even this plan failed: the "thing" went through the ceiling as quietly as possible, as if it were quite used to it.

So that's how it is, you see! When you think, "I'm now going to think about *nothing at all*; I'm going to allow myself to make totally free associations!" your thoughts will simply evaporate. You'll be too focused, your executive function will activate, and your daydreams will slip through your fingers. It's like sex—it's no good trying to have an orgasm; you'll only become stressed. You need to be present, but you can't concentrate too hard. It is zen. It is tao. It is knitting. It is yoga and meditation. When Siw Aduvill asks people during a yoga nidra class to be light and heavy (a box and a doll) at the same time, she is offering them a door to the DMN.

"I can fix your writer's block," Ida Jackson will regularly tell people asking for her advice. "Just get on the subway, then ride to the end of the line and back again, *without* looking at your cell phone! During this hour you'll have to confront all of your own insecurities and worries. You'll have to think about yourself, and whatever else you don't normally notice."

Research on the DMN and daydream mode shows that we need to spend quality time in our own heads. Up to 50 percent of our waking time should be spent in our own company—although since that's how our brains are equipped, we probably are anyway. But if our wandering thoughts are constantly interrupted by our cell phones, it's not surprising that it affects our well-being. Agnes Ravatn stopped checking her phone in order to become a better writer:

> Ideas and sentences, and not least observations, usually occur during small breaks, small pockets in time—when you're waiting for someone, sitting on the tram, going to work... And after I got an iPhone, whenever these pockets came to me I would pick up the phone. So I was throwing away the best opportunity I had for getting ideas, which is stupid when you want to make a living from writing. So for a while I trained myself to not reach for my cell phone every time I had a break.

In 2015, the journalist Manoush Zomorodi was hosting the podcast *Note to Self* and discovered that she was in a slightly dysfunctional relationship—with her phone. Seven years after getting her first smartphone, she suddenly realized what was missing in her life: boredom, daydreaming, and creativity. She started an extensive project aimed at helping people to put their cell phones aside and be more creative. It was called *Bored and Brilliant*, and later became a book. People would vividly describe the withdrawal symptoms they experienced after putting their phones away. One wrote how he had been diagnosed with ADHD as an adult, probably after suffering from it his entire life while finding ways of controlling it, but after too much smartphone use, it became full-blown. ADHD is connected with a failure of executive function.

When you're not constantly on your phone, you may also become more of a friend with yourself and have more ideas. However, you may find that you're bored more often, and you might even find your dark thoughts as well. If it's really true that the DMN is like a kind of thought soup (for lack of a better word), or a large, bright thing you can't quite grasp—as in Lewis Carroll's metaphor—then it's important what you put into it. Because what goes in will determine what comes out. If you fill your daydreams with Excel spreadsheets and years-old feuds and dark thoughts about yourself and gripes with the council about parking tickets and TV images of bloody conflicts and all the saddest things you can possibly think of—that might just be what your daydreams give you in return.

There is a Chinese fairy tale about a man who is given a magic pumpkin to grow, which, when ripe, will be able to conjure up any meal the man requests. While the other gardeners devotedly tend and water their own pumpkins, the lazy man puts the least possible amount of effort into cultivating his. Although neglected, the pumpkin grows just as big as the others, and when it finally ripens the man asks it to serve him the most delicious meal he can think of. But the pumpkin will serve nothing but porridge to the increasingly frustrated man. In the end, the head gardener comes over to the man and says, "This is what you get for cheating at your work. Now your pumpkin is cheating *you*!"

Like the pumpkin, how you manage your own inner world may not be visible to the outside world. But if you're not bored often enough or, at least, don't sit quietly often enough, or you don't fill your head with enough good things or acquire enough knowledge, if you don't discover new things and places and books and people, or if you don't learn enough about yourself and others, you'll find it difficult having any ideas or funny notions of your own. And yes, I understand—calling it the DMN makes it sound very scientific, but you've known all along that I'm talking about your *inner world*, your inner monologue, the garden of your mind. What you put into it and what you invest in it, what you plant in it, and how you look after yourself can impact your mental health and your creativity, both of which are seemingly dependent on all of these things. The pumpkin cannot work magic. The genius composing a symphony at the age of five is the exception; most people who make things have to put in good things—they have to practice and practice, and endure venturing into their own minds, before anything good will come out.

There is a proven correlation between abnormal activity in the DMN and a number of mental illnesses and ailments such as anxiety, depression, ADHD, epilepsy, and—yes—simply being unhappy or a little down in the mouth. It may be painful to admit, but some of Ficino's ridiculous theories about Saturn and melancholia might be close to the truth. When Goethe was troubled by depression, it coincided with his creativity in such a way that both the ideas and the ruminating may have occurred while he was in the DMN. Edvard Munch's creative force and the darkness in his work may also have originated from the same source and been part of the same system.

All this was going through my head as I lay in the water tank, where I paid money to try out "floating." For a second I'd forgotten I was there—you'd forgotten too, perhaps? Yes, I'm still splashing around in the plastic tank that's costing fifty dollars an hour!

Anyway, we're back, and just as I was thinking I couldn't bear any more, the closing music played. Exactly an hour had passed. It was a relief, and a bit of a surprise. I had come through it, and I'd done so without inflicting any pain on myself! At least I now fully understood the words of Oscar Wilde: "It is awfully hard work doing nothing." It was so incredibly boring! Not only that, but I suddenly realize—after paying and going back out on the street—that I may as well have lain totally still on the floor of my lounge for an hour, or in the bath perhaps, or even better: I could have gone for a swim in the fjord. So I decide to do all those things a little more often. It costs nothing to set your alarm and lie on the sofa or in the bath for an hour—an hour entirely alone in my own company, where anything can happen! Maybe I'll even start taking a nap.

We're now halfway through the forest, and I can see the end of the story: I now understand what's perhaps most important when it comes to being more creative. It is banal, yet very complicated. It is cutting edge-brain research and ancient wisdom at the same time.

When I was little, my father always told me that I should be bored. Was there some secret ancient wisdom to this, handed down from generation to generation, from fathers to their children over thousands of years? I still remember him at the cabin; my childhood summers during the 1980s were long and without direction, filled with all the life-threatening temptations of boredom. If Kierkegaard had seen me then, he would have been very worried for my future. My father would repair the cabin while I sat on the veranda staring into space, waiting for him to finish so that we could finally play badminton on the gravel track that snaked past the building. I loved wandering around in my own thoughts, but waiting was not my forte, and my younger siblings were just boring.

"I'm bored!" I'd complain. "What can I do?"

"Good!" my father would say. "It's good that you're bored!" And at that moment, the voices of all the other fathers, repeating exactly the same slogan, would echo all over the country. "It's good that you're bored!" they chimed optimistically at their world-weary children. I can't help feeling that they knew something we are in danger of forgetting, we who put an iPad in the hands of our children the moment there's any sign of them being under-stimulated.

I call my father, now a professor emeritus in philosophy, although not spending much of his retirement being bored. He still works very hard. He'll probably always have a Protestant work ethic deep down, no matter how much he pretends to enjoy life.

"When I was little, why did you say that it was good I was bored? Was it something you knew, something you'd learned? Where did you get it from?" I ask.

"I had to look after four children, in a tiny cabin. The weather was nearly always bad, and I had nothing to entertain you with. There was no wisdom in it; I was just frustrated," he admits.

I'm slightly disappointed with his reply. I'd hoped there might be some kind of hidden insight behind what had led us all into boredom's sweet embrace. However, with his words, my father had unwittingly opened the door to Alice's Wonderland for young Hilde. I got up and walked across the deck and into the cabin. And there, in front of the tiny homemade bookshelf, I stood for a while, thinking. Then I pulled out a book. It had a brown cover and a picture of a little girl on the front, below the words "Alice's Adventures in Wonderland by Lewis Carroll."

5 | How to Learn Less and Less

OR: I START SCHOOL AGAIN.

"Ten hours the first day,"
said the Mock Turtle: "nine the next, and so on."
"What a curious plan!" exclaimed Alice.
"That's the reason they're called lessons," the Gryphon
remarked: "because they lessen from day to day."

LEWIS CARROLL SPENT almost his entire life working in education as a math teacher at Oxford. School at the time was not aimed at the student's need for development or creativity, and most of those who received an education were poorly motivated upper-class boys who never reached academic stardom. Oxford was, as it is now, an elite institution, which meant that was where the upper class sent their teenage boys. Boys who would sooner be partying and hunting than learning mathematics. When Lewis Carroll worked at the university, the campus was always full of whining hunting dogs.

Carroll, however, had wound up there because he had a head for mathematics. He was the son of a priest and no typical upper-class boy. After studying at Oxford, he then became a math teacher and lived on the school grounds. He was one of eleven siblings—who eroded the family's wealth—and when his father died, Carroll had to be the breadwinner for those sisters who hadn't met a suitable husband, which was nearly all of them. Only one of his eight sisters married.

Life at the prestigious Christ Church in Oxford wasn't that bad for Carroll, who lived as a bachelor and worked as a church deacon in addition to being a math teacher. He also cultivated a rich social life, went on trips to the theater, and pursued hobbies like drawing and photography. His interest in the children he met was so all-encompassing that there was later speculation about whether he was a pedophile. Alice falling into the rabbit hole has been interpreted as a Freudian image of—yes, you guessed it—sex, of course. A few writers began questioning Carroll's sexuality as early as in the 1930s, with the first modern biography critical of Carroll published by Florence Becker Lennon in 1945. The speculation really picked up after World War II, when in 1947 the psychologist John Skinner wrote an essay about Carroll's problematic desires.

It is impossible for me to establish here what is true and false about things that happened in England in the latter half of the nineteenth century, and the debate rages on among Carroll's biographers. But it seems quite obvious that Lewis Carroll, who had eight younger siblings and was used to playing with and entertaining them at his rectory home, was especially used to interacting with children. He was clearly fond of their weird logic and penchant for nonsense. In this strict Victorian world, where the rules and social hierarchies were extremely rigid, it would have been quite liberating for a mathematically gifted man with a creative talent to get the opportunity to speak and write in a topsy-turvy manner—to someone who genuinely appreciated it.

"I cannot understand how anyone could be bored by little children," he told his younger colleague Arthur Girdlestone, who said that Carroll believed that his brain felt rested and refreshed after being around children. He despised small talk and the attempts of the chattering class to hide their emotions behind an "impenetrable mask of a conventional placidity." As we know, convention is the enemy of creativity, and children are the masters of the DMN and wild mind-wandering without purpose or meaning. Maybe Lewis Carroll had simply found the secret to his creativity in his little friends?

"Ah, you should have seen the ink there was in *my* days! (About the time of the battle of Waterloo: I was a soldier in that battle.) Why, you had only to pour a little of it on the paper, and it went on by itself!" he wrote in a strange and funny letter to a little boy, Bertie. Of course, he didn't write like this to his adult friends; self-writing ink would have been slightly more interesting for children than for Oxford professors. Among his colleagues, Lewis Carroll possibly felt like more of a stranger. He may have felt like someone who needed to hide his imagination.

Hiding your creativity throughout your school years is something the artist Sverre Malling knows a lot about. Where he grew up, in a tiny village just outside of Oslo in the 1970s and 1980s, art wasn't something that was encouraged; there were no drawing lessons at school, and so Malling did it in his spare time. He begged his parents to send him to a Steiner school (also known as a Waldorf school in the US), where pupils are educated according to the philosophy of Rudolf Steiner, with far greater focus on artistic and intellectual skills. But there was no Steiner school nearby, nor could Malling's parents understand why he wanted to devote his time to art. Nevertheless, from the age of sixteen Malling knew he wanted to be an artist, after watching a documentary about Salvador Dalí and the Surrealist movement.

"In Dalí, I found a carnivalesque joy at turning the world as we know it upside down. I saw this flamboyant and exceptional person who shamelessly defied the adult world's tiresome pressure to conform. Much of this visual world also touched upon the childlike imagination of the nursery. I was used to hearing adults telling children, 'Don't be creative, don't be weird!' Dalí, on the other hand, went to great lengths to be creative and was never afraid of standing out from the crowd. It became his trump card, his route to fame and iconic status. After watching the documentary about Dalí and the Surrealists, I knew it was all within reach, and that my past weaknesses could become future strengths. I could stop feeling ashamed," says Malling.

So Malling began dressing strangely. He would carry a cane and wear his great-grandfather's fur hat and coat to school. "And I remember the moment when I got on the school bus wearing this weird fur hat, and how everyone stared at me and laughed, and I knew there was no going back. It was a feeling of transgression, both frightening and liberating at the same time. Like many other people, I know how difficult it can be when so-called normality is upheld by 'ties and obligations.' It quickly becomes a very narrow world, full of expectations that, deep down, you actually want to renounce. So to wrench yourself free and challenge your surroundings is liberating."

At the age of seventeen, Malling made his debut at the National Art Exhibition (Høstutstillingen) after drawing relentlessly in his bedroom. This early breakthrough was of no thanks to his school—there was nobody around him applauding his strange activity. He then applied to several art schools, believing he would finally meet like-minded people. But when he started at the Oslo National Academy of the Arts at the age of twenty-two, he was instead struck by a feeling of shame, once again reminded that he was weird, an outsider, different from the others—this time because of Salvador Dalí, his great role model as a teenager.

"I soon realized that I needed to hide the fact that I liked Dalí, who was considered kitsch in the official art world. I was expected to distance myself from the fanciful and unbridled joy of drawing, and restrain my euphoric output. The 1990s saw the emergence of a new generation of young artists who were competent with theory and considered themselves academics, not whimsical bohemians. My professors would assure me that the idea of art having to be dissolute and imaginative was quite superficial. So I began to feel ashamed for having such puerile taste; my dispositions had once again been rejected," he says.

But once again, he turned this shame into a strength. Sverre Malling continued to draw, and is today a highly respected and critically acclaimed artist, despite how poorly understood he was by the school system. His obsession with drawing has made him famous. He is inspired by Theodor Kittelsen and Elsa Beskow, by the relationship between humans and nature, decline and decay, and by ideas of transience and exclusion. His best-known picture is a huge, meticulously detailed musk ox. While Norwegians see the musk ox as originating in Norway, it actually didn't (it was reintroduced to the country in the 1930s), so the drawing therefore challenges the perception of roots and origins, with its disturbingly melancholic and heavy figure, beautifully shaded by thousands of tiny pencil strokes. It looks frightening and primal and sad. Something quite dark hangs over the picture, as it does with many of Malling's motifs.

"Eventually I returned to the infantile, intuitive things rooted in childhood. Drawing is still a kind of childish escape route, a way of constantly dreaming my way out of the adult world's tedious journeyman trials. I've attempted to reach a world of boundless ideas. A kind of imaginary world seen through the eyes of a child. A way of seeing that is readily given up during adolescence, and which I'm trying to preserve in adulthood. A kind of extended room for movement, you could say, or perhaps more correctly: creativity. In other words, an opening that leads somewhere beyond the established framework. And perhaps it's a way of opposing classical education, by putting the mischievous and playful above respectable adult life," he says.

Malling has no rules for where his inspiration should come from—like Askeladden, he believes that everything has its use. The internet, Instagram, film, television, comics, computer games, books, posters, exhibitions, nightlife, museums, concerts . . . everything from fine art to pop culture.

"I want to move between these extremities, between the traditional and the playful. Between the cultured and the vulgar, the restrained and the overacted, the adult and the childish," he says.

"For me, it's about opposing the rehearsed and depicting new perspectives. Hopefully, my work can also offer space and some understanding to those who think differently and feel left out," he says, while leaning over a huge drawing of a dog that he is currently working on.

I buy a copy of Malling's book *Many a Blossom Shall Its Leaves Unfold*, which is full of pictures that remind me of Alice, since they are inspired by old book illustrations. Women holding large playing cards in their hands—could they be the Queen of Hearts' soldiers?

I take the book home and show it to my daughter, and afterwards we read *Alice's Adventures in Wonderland*. We meet the rebellious flamingos, the baby who turns into a pig, and the army of playing cards who attack Alice and lead her into the courtroom where she returns to her normal size, and my little girl thinks it's all entirely natural and logical. On our way to kindergarten the next day, she tells me what she's thinking: "Mama, what if we're living inside a giant tooth? What if the house we're living in is actually the tooth of a baby troll?"

Just why we might be living in the tooth of a baby troll was never explained, because two seconds later my daughter was striding into kindergarten, waving goodbye to me with a confident look on her face, and I thought about Lewis Carroll, who was so into things changing size—perhaps owing to personal experience with a special kind of temporal lobe epilepsy. When I collected my daughter later, she was holding yet another drawing of a flower. It was orange and yellow and green, her usual palette.

"Even kindergarten staff haven't learned how to work purposefully with creativity. If you let children draw whatever they want, without any restrictions or objective, they'll simply draw the same thing over and over again," says Erik Lerdahl, Norway's only professor of creativity. He works at Kristiania University College, mainly with students of design and communication.

Lerdahl teaches his students to develop their creativity by giving them clear parameters for a task. For example, he might get his students to mold a piece of clay while it's hidden under a black plastic sheet—to reduce their dependence on their sight—as a way of demonstrating how restrictions can be a source of creativity.

"One approach many creative people use is to push themselves into a self-imposed crisis, which will in turn force new solutions to appear. But that requires courage, and it can be quite an uncomfortable experience; it feels safe and comfortable to arrive at something we're already familiar with. On the other hand, to break with old patterns and familiar hypotheses is far more demanding. In Indian mythology, we have the god Shiva, who is the god of both destruction and creation. Destruction is part of the creative process. You have to be able to tear something down to build something new," says Lerdahl.

One thing Lerdahl has noticed is that his students take fewer risks than before. He believes social media and continual visibility have a restricting effect on the students, making them more concerned with internal and external critics. If you cannot make a fool of yourself or allow yourself to be open to absurd suggestions, you will rarely get exciting results. If you are too focused on finding a solution and being applauded, you're more likely to get a predictable result.

"Many of the creative methods I teach are aimed at forcing people's thoughts out of their fixed channels to create radical and unexpected results. If you're going to develop new solutions, you need to have a curious attitude, an ability to change your perspective, and an approach that allows you to play with accepted truths. In hindsight, a good solution will seem obvious, but getting there will rarely be so. You need to become good at surprising yourself," says the creativity professor.

"Some people have a naturally creative mindset; they're adventurous and just can't stop being creative. But I firmly believe that anyone can develop a creative muscle. Studies show that people thrive more in jobs where they are given the opportunity to be creative. Using your creativity is a source of a better life; it has an intrinsic value in itself."

Erik Lerdahl is also responsible for a separate course in creativity where 250 students learn creative methods. He says that his students will often sigh with relief when they realize that they can learn to be more creative. There's nothing hocus-pocus about it; it simply requires targeted training.

"The school system doesn't manage this responsibility well enough. Starting with kindergarten—where children are simply given crayons and somehow that's seen as creative, like that's sufficient," he says.

At the same time, various studies have shown that creative skills are now increasingly seen as important. In the future, we will have to be more, not less, creative. So it seems odd that children don't learn how to work creatively at school.

"These days I first have to unteach all the students' ingrained ways of thinking—before I can develop their creativity," he says.

Lerdahl is not alone in criticizing schools. The education system is constantly under attack, a political hot potato in any election campaign: everything we want for the future needs to find its way into the classroom and become a subject that can be taught. Students are constantly tested and evaluated, they start school earlier, and they also have to prepare for it earlier. In North America, children will begin school at four or five, often spending one or two years in preschool before that. My five-year-old soon has to put the trolls and baby trolls aside—and become measurable.

"Children are being trained to be researchers before they've even left kindergarten," says Markus Lindholm at the Norwegian Institute for Water Research (NIVA). Despite his passion for scientific research, he does not want preschool children to be learning scientific methods.

"They have to be curious first," he says.

"Children won't become scientists unless they are curious. All the goals set for them in kindergarten just give them information fatigue," says Lindholm, pointing out that curiosity is the very bedrock on which knowledge must be built.

If a child's curiosity isn't stimulated, they won't want to learn, nor will they have any good ideas.

"The workings of curiosity remain unknown. It's a quality that's always praised at Nobel banquets and political speeches of course, but no one is able to say how the school system either impedes or promotes it. Most people just assume that knowledge alone does the job. But in an age where all the information on earth is two clicks away, boredom levels are higher than ever, both in classrooms and auditoriums," he says.

Yes, boredom is important—as we now know—but children should also be curious and want to learn about the world around them, and if boredom descends on the classroom, we risk losing that curiosity altogether.

Lindholm was my teacher at school and the person who first aroused my curiosity for science. My junior-high and high-school years were spent at a Steiner school in west Oslo where the whole class would go walking in the nearby woods, looking for stones and plants that we then drew and described. Steiner schools are known for their emphasis on creativity, where pupils make their own textbooks, and teachers inspire wonder and curiosity through storytelling and exploration. In addition to his position at NIVA, Lindholm is also a professor at Rudolf Steiner University College in Oslo, where he trains Steiner teachers.

"But there are many problems with this type of school," he says. If the Norwegian school system is too goal oriented, the Steiner school is maybe too unstructured. "Many people say we don't need to learn that much nowadays, that revising for exams is unnecessary—now that we can google everything. But Google increasingly presupposes that you actually know something already. If you can't tell what's important from what's not important, how will you understand the results of an internet search? At a Steiner school, you can easily find yourself not doing enough studying or factual learning from books."

He also doubts whether the idea behind the Steiner school can work in the 2020s. "Parts of Steiner's ideas are superb examples of humanism, and that goes for many of his educational proposals. But I think it's difficult to get anything fruitful out of the esoteric and occult parts of anthroposophy."

The school's founder, Rudolf Steiner, had ideas about science that are no longer taken seriously, and the belief system he developed, anthroposophy, is often characterized as pure mysticism, and esoteric. But there was one thing that was very good about the school: we learned all sorts of different handicrafts. I learned carpentry, bookbinding(!), calligraphy, map surveying and cartography, and metalworking. We drew and painted in almost every class—and choral singing, theater, and music were a natural part of the school day. We learned to be excited, to own what we had learned with our hands and our bodies.

"You don't have to be an anthroposophist to use Steiner's teaching methods. It is the methodology and academic culture of curiosity that brings the learning environment together. The belief that knowledge becomes deeper, more binding, and personally anchored when the students also have to paint, model, and write their own stories," wrote Lindholm in a piece for the school's hundredth anniversary.

The fact that there are now three thousand of these schools and kindergartens around the world shows that there is a demand for a school that does something other than measure the pupil's results—a place the students can try different materials and expressions, within the security of the school's framework.

Lindholm is primarily concerned that a child's natural sense of wonder should be protected, and not become operationalized and goal oriented. We have to cherish that sense of awe if we are to find the researchers our future needs.

"To feed their curiosity, children need time to marvel at everything from rainbows and soap bubbles to crazy ants and earthworms. Humans aren't born into a childhood surrounded by things and objects, but into a world full of emotions, where they make friends and discover amazing things. And children must be allowed to stay like that until they have had enough. Many of them remain at this stage until well after they have started school—a place where snowmen, clouds, and creaking steps murmur wordless mysteries. Because this sense of wonder—which matures into a deep curiosity and enthusiasm for research—is rooted in that kind of childhood. Good research is not just about acumen; it's also about the ability to naively wonder."

• • • •

IN SCHOOLS AROUND the world, it's quite possible that the environment for curiosity and creativity is worsening. In 2011, a major study of the US school system concluded that creativity among schoolchildren had decreased. This was measured with the help of the Torrance Test.

Professor Kyung Hee Kim—who was a student of Ellis Paul Torrance, the man behind the test—looked at students from kindergarten to high school and found that creativity, at least the capacity for divergent thinking, had declined in the US between 1984 and 1990. And it has been going downhill ever since.

The professor of education studied data for several decades before she was able to determine that "children have become less emotionally expressive, less energetic, less talkative and verbally expressive, less humorous, less imaginative, less unconventional, less lively and passionate, less perceptive, less apt to connect seemingly irrelevant things, less synthesizing, and less likely to see things from a different angle." She called this "the creativity crisis."

This caused a lot of concern in the US—which is, after all, the motherland of entrepreneurship. If children were to stop being creative, innovation would be affected later on.

"Schools throughout the Western world are incredibly performance driven. And schools like that leave barely any room for creativity," says Espen Schaanning, a history of ideas professor at the University of Oslo.

Schaanning belongs to a research group that is studying the Nordic childhood. When I meet him, he has just published the book *Til alle barns beste?* (In the interests of all children?) and is working on another about critics of the American education system. Schools have become not only a storage institution, but also a heavily political project, a place where future hopes and dreams are realized through curricula and learning targets. The world's new "riches" are its children and young adults, who will be refined—from crazy chocolate-milk-juice inventors who live inside the teeth of baby trolls—into doctors and engineers and professors and researchers: people who can create a better future.

"So children are taught skills and knowledge, and then measured accordingly. But to be creative, you have to be allowed to follow your interests over time, and for as long as you want—if you do something for a long time, you'll get new ideas. The school day is now divided into small units where what students do is always decided by a timetable, and where the teacher measures each pupil's performance according to specific requirements. And the competency requirements are deliberately set so that only a minority of students will reach them," says Schaanning.

The school day is broken up in a way that gives students no time to dwell on anything—or any way of getting stuck and lost, as good researchers often do—yet it's these moments that foster creativity! And on top of that, you have the threat of grades, which are almost like a punishment, or an opportunity for you to feel defeated. Schaanning believes that in this system, there's very little room for trial and error, one of the basic premises for creative work.

"I've visited a lot of prisons and written about the prison system, and found a number of similarities between prison and school. In prison, there's a lot of creativity, but it's not necessarily the type of creativity we are looking for, and the same thing can be said about school: the type of creativity we see the most of is associated with evading the rules or the watchful eye of the prison guards or teacher—it's about finding the pockets that allow you to do what you want."

While the fifteen-year-old schoolgirl and environmentalist Greta Thunberg was inspiring hundreds of thousands of people to strike for a better climate, another thing became clear: many school systems were more concerned about tests and assignments than about creating engaged citizens equipped with a sense of critical awareness—with many school districts across Europe and North America refusing to excuse student absences.

"That's precisely what the school climate protesters did—they engaged themselves with society and the climate issues in numerous creative ways, but the school system responded by saying that this was something students would have to do in their spare time; it had nothing to do with school. But this is exactly what schools are supposed to help their students to achieve. It was a narrow use of bureaucratic rules to cover up what school should really be about," the professor believes.

Schaanning sees a clear trend: children are measured from the start, in a regime stretching all the way back to kindergarten. At school, many children who are unable to achieve the goals they have been set will get assistance from a special education teacher. So if a child is struggling with a subject, they have to do more of what they are struggling with. The special education teacher will give them extra homework, and both the homework and extra tuition will eat into their free time or their vacation.

"It's not my job, because I'm not making the curriculum, but if it was up to me I would get rid of all the grades and ranking for schoolchildren. And instead of dividing the school day and the school year into measurable units, couldn't we just decide on one goal that everyone might realistically achieve in tenth grade? A bit like a driving test, something where everyone can actually meet the requirements that are set, more or less. It would free up an incredible amount of time for the students. The things children are now evaluated for in school are of very little use to most of them. Today's schools are designed to create losers."

Children who were allowed to follow their whims in the past now have to attend a target-oriented school without any playtime. If Espen Schaanning could decide, school time would be filled with all kinds of activities, like working in a hospital, or learning how to chop wood, or helping out at a refugee center—and in addition, children would be taught basic language and math skills, the bare essentials necessary for tackling life after school. He believes that it is unimaginative of politicians to want to measure children in relation to international standards, and that that is the biggest obstacle to school becoming a friendlier place for creativity.

"I'm also worried that kindergarten will now be school-ified as well, so that even very young children will have to prepare for the school regime. And they shouldn't have to, on the contrary! Children between the ages of three and six have incredible imaginations, and come up with amazingly creative solutions when they encounter simple problems. The older the children get and the longer they are in school, the fewer original solutions they come up with. Their imaginations become one-sided."

The professor has three grandchildren and is worried, just as I am with my own daughter, about what's happening in their strange little heads when they start school. Because a system that creates losers must have its dark sides.

"We should start by asking ourselves: What makes a good life for children and young people today? To not begin there is to set up assumptions about what a good life is without even discussing it. Implicit in the stress we expose children to is this idea that it's good for them. We live in a very child-oriented culture, and we mean our children well. And that's a good starting point. But we also allow ourselves to be seduced by measurements and graphs. Can we really say that life is good for the children who don't meet these standards?"

Schaanning's own daily life is not goal oriented; it is not fragmented. For him, every single day starts with an hour-long walk in the forest, with no agenda, where his thoughts will fall into place and ideas will enter his head. And no one, except him, will evaluate how good they are. This is how it should be for our children too.

We don't need to be entirely pessimistic on behalf of children's creativity—it won't break down, of course. And there are still teachers and organizations and subjects that coax creativity from their students, when they're not already doing it themselves: playing without permission, evading the teacher's scrutiny, or allowing their minds to drift. There are still places at school where children can explore and express their creativity, assisted by people who work in creative professions. The author Guro Sibeko is one of them. Sibeko has performed slam poetry for many years and now travels to schools all over Norway with a sackful of poems, as part of a national program called the Cultural Schoolbag. In between the normal classes, Sibeko gives students the chance to immerse themselves in the art of writing using rhythm and rhyme.

When I meet Sibeko, she and a group of backing musicians are visiting a high school where a group of teenagers are soon hunched over their pens and papers discussing content and tone. The students learn about metaphors and music, and will compete against each other at the end of the class. This "competition" has no bearing on their grades, and the class itself is meant to be judged purely on the amount of creative joy it brings.

"Regular teaching isn't creative, and it doesn't allow you to be absorbed; you're always supposed to have something to do, or be busy with something. And I feel like the school system values logical-mathematical intelligence more than it does creative intelligence," says Raina Nilsen, eighteen, one of the slam poetry students there that day.

"We put them under positive stress with very short deadlines to stop them being so self-critical, and with so little time and at such a fast pace, they just have to make the best of it. Working with young adults is a dream, by the way; they do what they are told and come up with some great stuff," says Sibeko.

Sibeko is a qualified teacher with ten years of elementary school experience, and she has found that there's no room for poetry and music—or goofing around and getting lost—in the public schools she has taught at.

"Normally, there's very little room for creativity at school," she says. "Children who really want to do something creative have to find out what that is for themselves." Sometimes she will help them get started, though—like the student who learned to play guitar in just one week at school and, after a lot of practicing at home, became very good. But there is no time for resting or mind-wandering in the classroom. School is a workplace where one adult has to get almost thirty children to succeed every day. So it goes without saying that it's hard enough just recognizing every single student's needs, or making time for every single student's wishes and separate interests—let alone finding time for mind-wandering and daydreaming or sitting by the sea.

"Nurturing creativity is not what schools were built for, although some teachers do try. The most important thing I've done for students who seem to need something more is encourage them to do things at home. It's not something they can do at school. I'll give them homework like walking in the woods, and hope that some of them will like it and start doing it voluntarily, because I know that it's very good for their brain development. I try things like that," she says.

Sibeko's life is very much driven by her own inner motivation and the joy she gets from being creative.

"I can be quite creative as a teacher as well. Those students who are already creative are free to nurture it, unhindered. It's the other children I worry about," she says.

One person who has watched school develop over many years is Harriet Bjerrum Nielsen. She is a professor emeritus in gender studies. Starting in 1992, as part of her research on the Norwegian school system, Nielsen followed a class from their first day at elementary school until their final day at high school. Now she has once again entered a first-grade class as a researcher and is noticing huge differences since the previous time, the most obvious being how many more kids per teacher there are—and that Norwegian children now begin school a year earlier.

"Children have always gone to school, and always found it a bit of a drag—mainly because it restricts their freedom. Children today are too young when they start, and teachers despair at getting them to sit still in such a rigid system. And for today's children, this restriction of their freedom continues after school, with day care—you barely notice school's finished. They'll either be getting help with their schoolwork, or playing in the schoolyard, instead of going off to explore a forest or an exciting city."

There's a huge contrast between childhood thirty years ago and childhood today, according to the professor: "My son's childhood was spent going into the woods and playing in the neighborhood. For the generation born since 1990 or so, that free time has vanished, and their lives have been placed into an organized framework. Today's children come home to modern parents who engage them with music, theater, and sport. Their lives are constantly filled for them, like there's a lack of faith in their ability to fill their time themselves," she says.

She believes there is less time for daydreaming at school today than there was at the school she originally looked at. But then, as now, she had to ask herself why it took so long for the children to learn the most basic things in the classroom, and yet—in their free time, once they were interested in something, they could become totally absorbed, hungry for knowledge, and learn incredible things very quickly. Just ask an average eight-year-old about dinosaurs, and you'll find out how much they can learn.

"School is a constant battle between an adult trying to teach and a group of children who often have a completely different agenda. Some teachers manage to beat the social game playing, but it's a huge struggle against the intensity of children's culture. That's why it takes so long to teach them anything."

So this is the daily life of children today: they are measured by the results they get at school, and those measurements and tests consume more and more time. And this all happens while one of the most important concerns for children, the social game, is going on between them. Outside of school, their well-meaning parents initiate numerous activities for them, hoping to adequately prepare them for life after school. There are very few places for children to be bored anymore, where they can go into their own world, mull over their thoughts, wander aimlessly and have good—and not-so-good—ideas about seeing telephones or chocolate-milk-juice—things we now know are key to the development of a creative brain. This crucial game now has far less space.

"It's a massive experiment, and we don't know what will come out of it. It doesn't seem like anyone thinks that the rising levels of depression among young adults could be related to the so thoroughly organized lives of children and the pressure on them to achieve good grades. Today's parents are more ambitious for their children than ever, and at the same time, they're determined to make sure their children are constantly entertained. My son was allowed to do his own thing in peace; I had no idea that I was supposed to entertain him—it never occurred to me," says Nielsen.

In terms of development, most of what children learn is learned through play until they are seven years old, perhaps older. Play happens regardless of testing and measuring; it imitates social situations, and sets the stage for all kinds of strange "what if" thoughts.

"All the tests that children are put through restrict their creativity, their enthusiasm, and their ability to learn. What we actually end up with is three-day-knowledge, where the children learn something short-term in order to pass a test, and the motivation related to the actual content of the lesson disappears. Goal management takes what a child might be interested in and chops it up into small, measurable pieces—which are used as a substitute for context and conversation. This also restricts the teachers' own understanding of how to teach. Of course, formal thinking requires formalized schooling. But abstract thinking requires schooling too," she says.

Nielsen believes that modern schools are often affected by adult moralism. Somehow or other, the world of a child and how they develop is not accepted—children are expected to be kind and decent and behave nicely. But if they are to learn about their own complex emotional lives, they need to test out their aggression and grief and quarreling among themselves; if they don't practice being good people as children, then learning to regulate their emotional lives as an adult will, of course, be far more difficult. Nor can a child use their own boredom productively, because there's no place for boredom in modern childhood.

"Children are insanely bored at school, but they're not free to act on their boredom and use it creatively," she says.

Developmental psychologist Evalill Bølstad Karevold wants to change the school system to allow more room for both creative exploration and trial and error. The program she is planning is completely new. Her goal, in what is currently a pilot project, is to increase the emotional competence of teachers in the classroom—and teachers are keen to be involved, since so much of what happens in school is not research based, but politically controlled. They are currently running these pilot projects in elementary schools in Oslo and the surrounding area. Should the funding be approved, the TIK school intervention program, based on another program called Tuning in to Kids, will be evaluated in twenty Oslo schools in the first to fourth grades of elementary school.

"The reason we're doing this is because we want to create an environment where children can explore and cope with trying and failing. If they are in a safe learning environment, they will dare to become more creative," says Karevold.

This is how, by guiding teachers in emotional understanding and regulating, they can lay the foundations for good learning.

"This can then help improve the general classroom environment, particularly the emotional competence of the students," she says.

Karevold is not alone in worrying about the learning environment in schools, or in having ideas about how it can be changed for the better. Politicians, educators, psychologists, and other researchers all have their thoughts about how learning should happen. Right now, a lot of effort is being made to prevent boys from dropping out, because the statistics are clear: at school, girls are top of the class.

In 2017, Dr. Camilla Stoltenberg, head of the Norwegian Institute of Public Health, was asked by the then minister of education to take a look at Norway's schools, with regard to gender differences in particular.

"The committee's mandate wasn't to investigate creativity. It wasn't something we discussed," says Stoltenberg, who led the work on the report.

Her findings told an alarming story about the creative environment at school. And since very few experts, if any, have done similar research, the report is fairly unique.

I meet Stoltenberg on a bench in the center of Oslo, prior to her giving a lecture. We sit and watch stressed people rushing through the park, while discussing the stress that children are under at school.

"There were a number of specific problems we had to solve related to how, on average, boys perform worse than girls at school. The most important proposal we made, perhaps, was strengthening vocational training, and our most expensive proposal was full-time schooling," says Stoltenberg.

Full-time schooling—as opposed to the regular four-hour grade school day in Norway—is expensive. It involves professional teachers, on higher salaries, supervising the children for the entire day—from early morning until their parents collect them—instead of the day-care staff who normally take over when school finishes around 1 PM. But it does create the opportunity to fill the days with more movement, something that benefits both boys and girls—perhaps especially boys. The committee believed that schools that offer more activity will hopefully be able to bring the boys' results up to the same level as the girls. At the same time, the new curriculum proposed that students should immerse themselves far more in each individual subject, and therefore have more time to learn.

"What I was most surprised about was how many tests children have to take, and how few of these tests actually give us an overview of children's learning and development. They had to take every possible test: local, municipal, national, and international. We have suggested two tests, which are standardized and can be used by researchers. I believe that there should be less testing during a child's school years, and what tests there are should be of a higher quality, and standardized, so that they can also be used in research," says Stoltenberg.

The committee has also been involved in removing some of the focus on theory in high school, which contributes to as many as 30 percent of boys and 20 percent of girls failing to complete their five years there, which often means dropping out of the job market.

"High school never used to be quite so important, but for those who drop out today, it's far more difficult to get a job. So we have proposed 'an exploratory school year' based on the Danish model, a kind of college year between junior high and high school, to motivate students and increase their chances of completing high school. But this is just as much about establishing what a good life for children and young people actually is," says Stoltenberg.

There are several paradoxes in modern schooling. Society may be less brutal than it was in the past—there's more equality and inclusiveness—but the stress level among school students appears to have increased. Self-reported mental health problems have increased significantly. In 2010, a survey about the mental health of young Norwegians was initiated, and it showed that problems among girls—the winners at school—are constantly rising. In 2008, the number of antidepressants prescribed to older teenage girls was roughly double the amount prescribed to boys, and in the space of ten years, antidepressant use among girls increased by 59 percent. Of the 6,671 antidepressant users, 4,741 of them were girls. The increased use of antidepressants among teenagers between the ages of fifteen and nineteen has been startling: in ten years, the total consumption rate has increased by 48 percent.

Mental disorders have also been recently increasing among boys. This rise in diagnoses could be due to young people being freer to talk about their mental health, or it could be due to greater pressure to achieve and the stress of worrying about not succeeding.

"We don't know enough about why so many young people develop mental disorders, or how they're related to pressure at school," says Stoltenberg.

As director of the Norwegian Institute of Public Health, Stoltenberg's job involves monitoring the state of the nation's health. The stress felt by the nation's youth, which is being measured by the researchers, doesn't quite relate to the broader society we live in, so much of which functions better than it did before. But what's missing entirely is indirection. If it's so important for children and teenagers to consolidate their memories, imagine the future, work on their relationships with themselves and others, daydream about other worlds, and explore their inner universe—why isn't there more space for it in their everyday lives? Can these pockets of nothingness and directionlessness, so necessary for nurturing both creativity and well-being, be found at school?

"That question probably applies to everyone. When the debate about our report was at its most intense, I was in the city of Bergen. I was walking while talking on the phone, and without realizing, I wandered into an unfamiliar part of town. Then my cell phone battery died, and I suddenly found myself lost and unable to use the GPS. But it was actually very nice. No one could get hold of me. Even I didn't know where I was. It reminded me of the time I traveled to central Europe when I was fifteen. It was the ninth of January, and I didn't call my parents until the fifth of February, because then it was my birthday. To be that unavailable is a rarity now. These days we are continually available. Children too. And it has a profound effect on us," says Stoltenberg.

For the duration of our sixty minutes on the park bench, neither of us has checked our cell phones. To a passerby, it could almost seem like Stoltenberg, the woman who postponed our interview several times due to pressing engagements at the Institute of Public Health, has plenty of time on her hands. She stands up, ready to mount her bike and ride away to the next conversation about school.

"You seem very relaxed for someone so constantly busy?" I remark.

"Ha, yes, I hope that's true," she says. "I like to think that if I'm in the middle of a storm, I'm in the eye of the storm, where it's quiet."

What's it like to be at elementary school today? After talking to so many people with opinions about school life, I feel like I really should try it for myself. So I decide to become a student again and return to school, to experience firsthand what it's like to be the subject of an ongoing political tug-of-war.

Before Steiner school, I attended an elementary school located in the idyllic and fairly affluent west Oslo neighborhood I grew up in. It was to this school I was going, to spend a whole day in third grade. Families in this part of town are well educated, they go hiking in the nearby forest, and they have a life expectancy ten years longer than people on Oslo's east side, where I live now. (I never should have moved; now I'll probably die at the age of seventy from some stress-related illness.)

When I was growing up, the hundred-year-old villas and town houses were mainly occupied by academics, people who worked either at the university or at the nearby University Hospital. Back then, these homes were easily affordable for a not-so-well-to-do family of six, like ours. Now the people living here belong to the upper stratum of the cultural and academic world.

I cycle past beautiful flourishing gardens that make me think of Virginia Woolf's secret paradise in Sussex. Here, people go around pruning their rosebushes and apple trees in the middle of the day, or they sit outside the local bakery enjoying coffee and cakes, without displaying the tiniest hint of stress about pending tax returns, car safety checks, children's birthdays, or scout camps.

But my pulse quickens as I approach the pink 1920s building sitting majestically among the gardens. My time at this school was actually spent feeling extremely lonely and inadequate. I would often stay indoors and read to myself during recess, because I had no friends, and the social game playing in the schoolyard was too exhausting. Those who now describe me as outgoing and bubbly would never have recognized the silent, introverted little girl I was at this school. In third grade—the same class I am about to return to—a decision was made for me to have extra supervision for English. I can clearly remember the humiliation, a humiliation I've managed to hide behind a master's degree on Shakespeare and a British husband who I practice my English with every single day. *I used to be so bad at English,* I'll think in amazement when a particularly melodious English word rolls off my tongue. *I really shouldn't be able to do this.* Will they remember all this when I come back, thirty-five years later?

The school looks almost the same as it did then, except now everything is far better maintained and there's brand-new playground equipment outside, built according to all the proper safety requirements with a soft rubber surface. The forest of stinging nettles that was once an impenetrable and frightening jungle behind the school is now gone—and the dark and creepy attic, which I sometimes tried walking through during the lonely recesses, has now been converted into a bright and cheerful admin room for the school's forty or so teachers.

It's up in the admin room that I'm met by Aria, a young teacher with an intelligent demeanor, who seems remarkably calm considering she's about to take charge of a huge gang of eight-year-olds. It has to be a good sign. Aria has been teaching for five years, three years at this school, and has a head for math, she tells me—she wanted to become either a teacher or an engineer, but since there was a waiting list for engineering school, she opted for teaching.

As the lesson is about to begin downstairs, the students quietly enter the classroom and sit around four large tables in the middle of the room, and at desks along the wall where there are iPads. They learn in stations, which means that every time the children switch tables, they change activity. This happens every fifteen minutes. The children knit at one table, they solve tasks on the iPads at another, they practice story writing at the third table, and they play a memory game at the fourth table that involves retrieving words from another room and writing them down. The classroom walls are covered in posters reminding the children to be kind to each other, showing how they can resolve conflicts; one shows traffic lights for regulating voice levels (from red, which is totally silent, to green for talking, via yellow for whispering). There are drawings of the Bronze Age, a subject they have recently studied; photos of the students labeled with their dates of birth; and in the window, a row of sunflower shoots, twenty-five in total, one for each student.

There are also two posters that I don't notice at first: one is about "wolf language," the other about "giraffe language." Wolf language is about judging right and wrong, criticizing and offending, and a wolf will also make demands and give orders. It sounds very much like people who are in the sympathetic system—stressed, agitated, and focused, ego-driven and snappy.

"We've called it 'giraffe language' because it takes so long to travel up through your neck and out of your mouth," the teacher explains to me.

It is parasympathetic language, about openly expressing your feelings and showing respect, praise, and support for others. It makes me think of the award-winning journalist and teacher Esther Wojcicki, whose three daughters, all with very creative high-flying careers, are proof of her abilities: one became the senior vice president at Google before moving on to become CEO of YouTube, another is a pediatrician and professor at the University of California, San Francisco, and the third is CEO of the DNA company 23andMe. Esther Wojcicki has written the book *How to Raise Successful People*, which has been a success for the author too, and her exceptional family has appeared in several major newspapers.

"Without really intending to, I found I'd started a debate about how we should be raising our kids and how to make education both relevant and useful," Wojcicki wrote in an article for *Time* magazine, believing that her method could be an antidote to discipline problems, power struggles, and fear among parents.

"We are in a crisis of trust the world over. Parents are afraid, and that makes our children afraid—to be who they are, to take risks, to stand up against injustice. Trust has to start with us. When we're confident in the choices we make as parents, we can then trust our children to take important and necessary steps toward empowerment and independence," says Wojcicki.

Wojcicki calls her values "TRICK": trust, respect, independence, collaboration, and kindness. "It's important to show our kids that the most exciting and rewarding thing you can do is to make someone else's life better," she writes.

She is not the only one trying to start this revolution. Many people, even in a place as competitive as the United States, are against competitive schools and the pressure to get better grades. People such as Michelle Obama, who made the following comments at a talk she gave in Oslo, when promoting her book *Becoming*.

"I think it's wrong that we have a school system that tests young children. I remember personally how as a first grader I was tested on something I could actually do, but couldn't do in that situation, I was asked to spell the word 'white,' and I just couldn't do it. I froze. Testing children, as much as they do at school today, robs them of their commitment and passion," said the former first lady.

Her opinion sums up what memory research has also shown: you cannot be certain that knowledge will come to you when you need it. In a test situation, stress alone can make you forget something you might normally be able to remember. Passion and commitment are what make you remember and retain knowledge, and enable you to use it creatively. If you cease to be curious and engaged, this knowledge will not become a part of you—it will be something you just retain and then forget later.

Of course, what these two highly respected women say conflicts with a centuries-old principle of child-rearing: children should be disciplined, shaped, and tested—preferably until they buckle under. How else are we to know if they have done something properly? Despite several positive modern reforms, however, there is more achievement pressure in school now than ever before. And if our children's lives are governed by stress and fear of failure, will they be able to think strangely and take the risks necessary for creating a new and different future? At the elementary school I'm visiting, on this summer day in this very well-resourced neighborhood, I'm on the lookout for daydreaming, alternative thinking, and a willingness to take risks among the third-grade kids.

I look around the relatively orderly and well-behaved class. Some of them writhe in their seats. One little girl kicks off her shoe and limps impassively across the room to retrieve it. I see the face of my own child in her soft features, and my friends' children in the faces of the neatly presented boys. Humming to themselves while they knit, squirming restlessly and joking—so much future in these hands and heads! After an hour of knitting and writing, Aria asks the whole class to stop and listen.

"Now we're going to play dodgeball," she says, dividing the class into two teams.

"They're all so restless now, I have to take them outside and let them run around a little. It's good for their concentration," she says as we go out to the back of the school.

But this isn't in the schedule—it's something the teacher has decided to do, and it backs up what many people think about school: that children need to be more active.

When the game is over, it's back to the classroom for more knitting and writing. I see that nearly all the boys are now reading in earnest—titles like *Diary of a Wimpy Kid, The World's Coolest Gang,* and *Demon Dentist.* The little eight- and nine-year-olds sit hunched over their books; they fiddle and flounder, making exceptionally slow work of the brightly colored patches they're knitting, like they're slowly pulling a dandelion from the ground with every stitch they pull up. But they keep going.

"My grandmother's a real pro at this," says one of the boys as he lets rip with the knitting needles, full of determination.

"I know there's often a problem with boys not reading, but that's not the case in my class. The boys actually read more than the girls here," Aria tells me.

After recess, Aria keeps up the students' physical activity and takes them out in the rain to play another ball game involving English words. Whenever a child is hit by the ball, they have to go to the teacher, who will show the child a note with a drawing on it, and the child has to say what the drawing is, in English, before they can run back onto the field again. If the student cannot say what it is, he or she has to do ten jumps and try another word. The children throw the ball at each other and shout "cucumber" and "watermelon" and jump and play and laugh and eventually return to their seats, warm and happy.

"We know, of course, that children at this age only have an attention span of about fifteen minutes. So I adjust my teaching accordingly, and focus on each subject for about a quarter of an hour," the teacher explains to me after school is over.

"I try to keep them focused and awake throughout the day. I don't want them zoning out."

"But why not?" I say. "Have you not heard of daydream mode?"

She looks genuinely surprised. It's clear that she has never considered that it might be nice for children to be alone with their own thoughts now and then, that it's there they can consolidate what they have learned, work on their relationships with themselves and others, envision the future, and have creative ideas. It's not that she's a bad teacher; she is one of the good ones. It's just not something teachers learn about before entering the school system.

On the way home from my reunion with elementary school, I went to pick up my daughter from kindergarten. But on the subway ride there, I took the opportunity to digitally unregister her from the dance class I'd been taking her to for the past year. Then, when we arrived home, I confiscated the iPad she loved so much. And there I stood, with my daughter glaring at me from the other side of the living room. She was furious, demanding the return of her iPad. But I decided to endure all the anger and pleading this time.

"It's good for you to be bored!" I tried telling her, but she seemed remarkably disinterested in hearing about the DMN research. I looked at the clock, counting the seconds as our cold war continued and the seconds became minutes. Then, she abruptly lost interest in the conflict, stopped whining and begging, and walked quietly into her bedroom. Not long after, I heard her talking loudly to herself. A great drama was about to unfold in there. One of her dolls was clearly in mortal danger and would die unless one of the stuffed animals rescued her. I sat on the sofa and grabbed my cell phone. Then I stopped, put the phone down, and sat there in complete silence, letting my thoughts fly.

6 | The Art of Painting White Roses Red

OR: I QUIT MY JOB.

"They're dreadfully fond of beheading people here; the great wonder is, that there's any one left alive!"

"EXCUSE ME," I said. "I haven't been here before. Where do we hang up the coats?"

My daughter had actually been going to her new kindergarten for three months before I had the opportunity to drop her off or pick her up. When I finally did, I had no idea where her things were kept or what the other children and parents were called. I had been working from 6 AM until past midnight some days. There were days I didn't say good night to her; there were days when I fell asleep in the middle of a sentence, on the hard leather sofa in the living room, while the darkness outside pressed against the windows.

At school, when children are struggling, they are just expected to work harder and use more of their free time studying. Which is just how I felt. I had just started a new job as an editor at a publishing house and had a concussion that made me feel continually exhausted. Not only that, but I was employed on a temporary contract, which basically meant that I had to work unpaid overtime—yes, it seemed like the job was deliberately meant to take over my entire life. My boss let me know early on that I was behind schedule and had to make up for lost time, that I risked losing everything. She thought I was performing badly, that I was a disappointment. So from that point on, I thought about work all the time.

Maybe there was a reason for me performing so badly? Perhaps the concussion had left my brain damaged? My ears were always ringing and I had to lie down constantly; I would become tired and unfocused—suddenly and with no warning. And that's precisely why I thought that I needed to work harder. To impress my boss. I had to rectify any impression I had given of being lazy, or even a bit stupid. The Queen of Hearts sat on my back and shouted insults. So I decided to work during the weekends as well. And I went in to work during Christmas, of course—*no problem, it's the very least I can do!*

"We have to cancel the Christmas holiday," I told my husband. Instead, he gave me tickets to a children's play that I could go and see with my little girl, who I barely saw anymore. Finally, we could at least experience something together. We put on red dresses and went hand in hand to the little theater two days after Christmas, while the biting cold froze our cheeks and the snow creaked under our shoes. But as the show was about to start, I received a text message from my boss. I should have known. I had made a mistake. Again. I got a sinking feeling in my stomach, and before even thinking about it, I jumped into a taxi and raced to work to correct as much as I could. It was the middle of the Christmas holidays, and I had left my little red-dressed girl, so full of expectation, at the theater with my husband. In the taxi, I fidgeted nervously with my hands, pulled my mittens off, and bit my nails anxiously. Still, I knew it was all futile, that whatever I did now wouldn't help. For the first time in my life, I was one of those people who failed to meet the mark. I was a loser. I had lost.

Throughout this whole period, I stopped sleeping, I stopped exercising, my palms sweated day and night, I had a constant knot in my stomach, and I lost my appetite. My immune system weakened, my pupils were dilated, my digestive system behaved strangely, and I stopped laughing altogether. I stopped listening to music, meeting friends; I stopped leaping spontaneously to my feet to dance, or cracking jokes, or trying to notice unusual things about people I saw on the bus (like I usually do). Because, yes, you guessed it: I had activated the *sympathicus*, the body's stress response and lifesaving state of emergency. I felt like I was in a major crisis that went on and on, for days and then months. And when you are in such a state, it's impossible to be creative, because everything has become life-threatening, critical, hanging in the balance. A dull sense of purposeful gravity descends when you realize you're constantly three steps behind, always catching up. When I found myself in this situation, I avoided wandering around in my thoughts, because, of course, it wasn't efficient enough. I didn't dare try anything new, because failure was no longer an option. It's impossible to have anything beyond conventional thoughts in your head when you know you're running along the edge of a precipice—and just one tiny slip will send you plunging to your death. But that's how I ran. For a long time.

The journalist Johann Hari wondered why anxiety disorders and cases of depression are on the increase in the West, despite the fact that we are better off materially than ever. In his best-selling book *Lost Connections*, Hari found that, among other things, one of the causes of depression is work. In general, work is not a place for expressing ourselves creatively; it's somewhere we live out the pointlessness of being there. In 2011 and 2012,

Gallup spoke to millions of workers across 142 countries. The worldwide survey showed that only 13 percent of those asked were interested in their work, 63 percent were not the slightest bit interested, and 24 percent were seriously unhappy at work. This basically means that most people, as many as 87 percent of all employees, do not feel that they belong in their workplace—and, as a result, have no good ideas or creative thoughts with regard to work. Even in safe and secure Norway, a country abundant with rules and resources, the National Institute of Occupational Health found that an astonishing 135,000 people, out of three million workers, were considering suicide as a result of bullying at work. Some 123,000 employees experienced unwanted sexual attention more than once a month at work—that's 4.7 percent of the country's workforce.

In the USA, 15.5 million people take antidepressants. But is that the solution? Is it possible to just medicate the problem away?

"We need to feel we belong to a group; we need to feel we have a stable future; we need to feel that we are valued; we need to feel we have meaning and purpose in our lives," writes Johann Hari about the epidemic of depression in the West.

In other words: we need to feel like we're in *parasympathicus* more often. Stress makes us feel lonely and scared; *parasympathicus* makes us start friendships—it helps us create the time and space to think our own thoughts, to daydream and be creative, to laugh and play.

"When I began researching stress, I found that it was the source of much of the hopelessness in the world," says Anne Gunn Halvorsen, who has written the Norwegian book *Stress og korleis leve med det* (Stress and how to live with it).

"Stress causes depression in adolescents and divorce among adults, and is one of the reasons for serious and long-term sick leave because it leads to cardiovascular disease, skeletal and muscular disorders, depression, and anxiety—basically, all the main reasons for long-term sick leave. It is a structural problem, and yet the finger gets pointed at every one of us individually, as though it were somehow our own personal responsibility! The biggest problem with stress is the absence of any political will to discuss it," she says.

Halvorsen believes that our constant pushing for more growth makes it impossible to discuss things like working less, for example. Six-hour days or shorter working weeks are not on the table during political negotiations.

"But we have to stop and ask ourselves: What is it we actually need? Stress costs society enormously. We each pay a price as an individual—or individual business—and that makes it hard to see that it applies to all of us, collectively," she says.

As a mother with young children, Halvorsen knows that having both parents working full-time is one of the main sources of stress. It is, of course, a relatively new concept that has never been tried before, and has radically changed the family unit over the past forty to fifty years.

"The nuclear family is way too fragile. Why do we detach ourselves from our extended families? We end up so alone! And when we work so much and then add children to the equation, it becomes very difficult to live collectively. We have so much, and there's so much to lose. In this society, many people are afraid to take any risks," she says.

Halvorsen thinks employers should have more faith in their employees' ability to manage their own workday. It makes the staff healthier, less stressed, and more creative. A large number of employees just surf the internet and pretend to work when they've completed their work for the day, instead of going home and using their energy on something else. Of course, this too makes them stressed. Halvorsen has noticed how stress makes her own life feel very small—it becomes impossible to picture anything fun or exciting in the future.

"The world has become a ruthless place for today's youth. Their whole lives are dependent on them passing that specific exam, getting into the housing market, and finding a job. All of this just limits their imaginations. They never have time to look up. And the same thing is happening in the workplace, where temporary posts have become increasingly common and employees feel even more stressed. It makes it impossible to think long-term," she says.

Someone who is stressed will be more prone to falling for distractions like TV—or any other type of screen-based entertainment—and eating unhealthily. Stress makes you think short-term thoughts; it's all part of the nature of stress, to not make long-term plans.

"Since writing a book about stress, I've started meditating regularly, for twenty-five minutes on the living room floor, and it energizes me so much more than all the distractions. It's what we did before: we'd take an afternoon nap. Whatever happened to the afternoon nap?" she says.

Halvorsen, a journalist herself, is skeptical toward much of today's reporting. She is also wary of consumerism. These two powerful forces simply drive the pace of society even faster: you could lose everything in a heartbeat, maybe to terrorism or to the climate crisis—you read it all in the headlines. And that actually makes stress your worst enemy, because you *can* lose everything in a heartbeat—if the stress, unhealthy lifestyle, fear, and hassle give you a heart attack. You can end up with a worse life—a less creative life—than you really need to have.

• • • •

WASIM ZAHID TREATS some of stress's most fatal consequences. Ironically, his own workday is also full of stress, albeit in a more structured form.

"I'm not thinking creatively when someone has a heart attack," says the cardiologist.

In an emergency, Zahid is not meant to be thinking creatively at all. It's all about well-drilled procedures. When you have a cardiac arrest, you definitely don't want your doctor to think creatively; you want that person to get your heart started as quickly as possible. This means that Zahid has to be totally alert and let the routines he's practiced do the job. When a heart attack patient arrives in the emergency room, there's no time for daydreaming.

On the other hand, Zahid does have an extremely meaningful job. He is among the 13 percent of employees who find their job really rewarding and deeply meaningful. "Doctor" is what many school highfliers want to be when they're clutching their exam results, ready to choose a career path—but it doesn't necessarily mean you have a creative life.

Nevertheless, Dr. Zahid cannot stop himself. His good ideas come only at the worst possible moments and when he least expects them to, sprouting up like dandelion shoots through a crack in the pavement. Not only has he written a book about the heart, he also has a newspaper column. And now, when he's not saving lives, he also has his own YouTube channel—because saving lives isn't enough! He wants to share his knowledge, and use all his creative energy to make that happen. And when there's no time during the day for his ideas, they come to him when he finally has time to relax, in bed.

"It's mostly when I'm in bed. My thoughts will start swirling. I'll be unable to sleep, because my head is so full of ideas," says the doctor.

Zahid's case shows that our brains are made to daydream and come up with good ideas, whether our jobs allow it or not. As with cardiologists, there are some professions where you would think daydreaming on the job would be almost impossible, like being a prime minister, or a CEO. My boss's boss's boss, Kristin Skogen Lund, for example. Does she ever have time for daydreaming? Lund has five thousand employees beneath her and is responsible for a turnover of billions.

"Yes, I find pockets of time, and I hope everyone tries to do that," says Skogen Lund, CEO of Schibsted—the large Scandinavian media company that owns the newspaper *Aftenposten*, where I've written regularly as a freelance book critic for the last six years.

When I finally get her on the phone, she's on a business trip to Paris.

"Traveling, like I am now, is one such pocket—it's a break from daily life that allows me to think differently. I get creative mostly from talking to people; I'm very outgoing and always learn new things from those around me. But there's a limit, of course, as to how far I can reach into the organization. Those I'm most in touch with are the managers just below me," she says, before assuring me that she wants to create a culture of listening to each other in the workplace.

Skogen Lund has previously been vice president, president, and CEO of the Confederation of Norwegian Enterprise, and has a unique insight into working life from an employer's perspective. She has also been executive vice president of telecommunications company Telenor, has held numerous board positions, and has been named Norway's most powerful woman, twice.

"But would your methods trickle all the way down to me? Would I somehow be able to tell that you're the boss of the newspaper I am writing for?" I ask.

"Not immediately, but you'll notice in the end. I believe in a combination of freedom and clear direction. As a manager, you have to lead the way, but you can offer a lot of freedom within any framework you set. I believe that with the various businesses, as much as possible should be decided as closely as possible by those who know them best. Leadership is ultimately about performing through others. It means offering freedom and motivation—and I believe this is something that reverberates throughout the organization."

"Could you envision introducing a six-hour day at Schibsted, if you knew it would increase creativity? Or a four-day week?

Or the Google model, where employees are allowed to manage 20 percent of their working hours as they see fit?"

"Well, not everyone at Schibsted works in the same way now. Some work shifts, others have fixed hours, and others have more freedom to choose their own working hours. For example, it would be hard to get everyone in the newspaper editorial office to work at the same time—when news has to be monitored around the clock—or to try to establish working hours that suited both newspaper couriers and office staff," she says.

"Having said that, I'm a great believer in devoting time to creativity. So it's also part of Schibsted's strategy for all its employees to be able to spend part of their time being innovative and creative, in addition to doing what they already do. But I have more faith in us being creative within our existing working hours than by working less."

So Schibsted will definitely *not* be doing what Stefan Sagmeister did to protect his own creativity. Sagmeister is one of the world's leading designers; his clients have included the Rolling Stones, Lou Reed, the Guggenheim Museum, and HBO. His studio was originally located at home, a two-story apartment in New York where he lived on one floor and worked on the other. To make sure he didn't suffer designer burnout, he made a rule: to never work on weekends and never work after seven in the evening, despite the fact that he loved designing. For a young superstar designer to decide that his office would *not* work around the clock was highly unusual. But he clearly knew what would happen if those limits were not set: burnout by the time he was thirty-five. Over the years, he had noticed that office routines could be a threat to his creativity.

"I saw how design students would spend a lot of time on pure experimentation, and it made me envious," he says.

Then Sagmeister had a radical idea: his super-trendy design agency would close for one year. After warning all his customers a year in advance, he locked the office door for the last time. The designers working there were then expected to go out into the world and gather inspiration. It was a sabbatical year. And the agency still does it, every seven years.

There is an ancient Jewish custom described in the Torah where every seven years, the fields must lie fallow—and every fifty years, all debts and guilt must be cancelled. The teachers at Steiner schools have sabbatical years too, every seven years. And it's a growing trend—just google "sabbatical" and you'll find testimonies from recent converts about how good theirs has been. It's also not unusual for top chefs to do what Sagmeister did with his agency. One of the world's best restaurants, Noma, in Copenhagen, was recently closed for a year and then reopened—revitalized and full of new ideas.

"The first time I took the sabbatical, I quickly found out that my initial desire to conduct that year without a plan was ill-fated. I ended up plotting everything I'd dreamed of doing while working full-time into my calendar, sorting it by importance into three-, two-, and one-hour segments, and wound up with a schedule, just like elementary school," Sagmeister tells me.

Despite this year being a little overplanned, it was a success. The sabbaticals that followed were not quite as strictly organized.

"I had all sorts of fears before the first year—that we would lose all our clients, that we would be forgotten, that we'd have to start from scratch, that people would see it as unprofessional—none of which turned out to be the case. By the time of our second and third sabbatical years, I'd almost stopped worrying entirely. Now I get lots of people saying how envious they are when I talk about the sabbatical, and many of them say they would like to do the same."

What's special about the sabbatical year is that it works as a source of creativity for everyone at the office. All of the interesting ideas they are working on come out of the period when they're not at work. But what do we mean by "not at *work*?"

"Hour-wise, I'm probably busier during this sabbatical year than I am normally, when I actually have *customers*. And it's not about partying. I can party and have fun anytime; I don't need an experimental year for that! Ultimately, the best outcome of the sabbaticals is that they make sure that my work remains a calling, and doesn't deteriorate into a career or a job. I think that anyone whose job description includes 'thinking' or coming up with ideas will benefit from this tremendously," says Sagmeister.

So he spends his time traveling, meditating, and experimenting—and seven great years come flooding out.

But there are other ways of creating pockets in time too, places where daydreaming and creativity can occur in the workplace. The Danish IT company IIH Nordic has introduced a twenty-five-minute break during the workday, during which employees cannot be disturbed by email, phone, or their colleagues. A small red lamp will indicate that they must be left alone. The company also limits meeting lengths to forty-five minutes. Not only that, they have introduced a four-day working week for all employees, while paying a five-day salary. As a result, the company has actually increased its productivity, despite the fact that its employees registered fourteen thousand fewer work hours than the previous year, when they worked a five-day week. On top of that, the rate of sick leave has decreased. Great Place to Work recently named IIH Nordic the best small IT company in Denmark.

"I think the four-day week will give a competitive advantage to employers, and is something that will spread. First in the private sector, then the public sector," says Dennis Nørmark.

Nørmark is a social anthropologist and, along with the philosopher Anders Fogh Jensen, he has written the book *Pseudowork*, which caused a lot of debate about the culture of work in Denmark when it was published in 2018.

"It started when I heard the expression 'bullshit job,' and I suddenly realized it was what I had. I had a bullshit job. I would say that I was going home early to continue working there. And when I arrived home, knowing full well that I had nothing to do,

I no longer had an audience, so there was no point in continuing to pretend I was working," says Nørmark, who worked as a consultant and partner in the company Living Institute, where he used his expertise in social anthropology to train companies to understand cultural differences.

He quit his job in 2014, and has been his own boss ever since. Now he is a well-known speaker, author, and writer.

"When you're writing a book, you can easily work a sixty-hour week without being stressed; you lose any sense of time and place because you're doing something meaningful. A great deal of the stress people experience comes from doing jobs they don't really think are meaningful," he says.

When he and his coauthor researched the culture of work, they found that working fifteen hours a week really should be more than adequate! Our core tasks can be done in that time, they say.

"By pseudowork, we mean creating the illusion of being significant, and that means appearing to be very busy. You'll be rewarded for filling out a form, but not for doing something unexpected. Systems like this make it impossible to be creative or generate anything new," says Nørmark.

Nørmark has very little faith in doing the occasional creative workshop; what's needed for an innovative work life is radical change. He believes we need to move from filling our days with things we don't need to do to spending our time on things we think are meaningful for the good of the community, and then go home. Even if it's the middle of the day.

"The system we have right now is insane! Work life is absolutely insane!" exclaims Dennis Nørmark.

The philosopher and coauthor Anders Fogh Jensen believes we need to value free time and idleness far more, and remove the traditional link between working hours and wages. He and Nørmark are on opposing sides politically, and Jensen is happy to be self-critical on behalf of the left. At the beginning of the twentieth century, a seemingly unbreakable bond was created between working hours and wages. The unions and employers agreed on the eight-hour day and negotiated wages on that basis. However, what they were negotiating about at the time was hard physical work in factories and heavy industry—and the eight-hour day was a step forward for workers who often worked long days and were engaged in tough manual labor. Since then, our lives have changed, and most people in the West no longer sit on assembly lines, but in front of computers. And although computers make us work faster than ever, the eight-hour day remains the standard. Twenty-five years ago, when I began my career as a journalist, we couldn't use the internet for research, since what little there was online back then was total garbage, and there were no good search engines. We were always dependent on browsing through a paper encyclopedia. All communication with sources went via a landline or by fax. Writing a news story then took me twice as long as it did ten years later, when the internet, cell phones, and email had become commonplace. But that didn't mean I worked shorter days. I just started doing more. That is now the standard in today's workplace. And if we're not working more, we just pretend we are.

"But it's not the amount of time you spend on something that counts. No one asks me how many hours I've spent preparing when I give a lecture—I'm not paid for those hours, but for what I deliver. So much of our identity is associated with work and working hours, and we still live by a Protestant work ethic. Most people have this idea that if you're constantly busy, then you're a very important person," says Jensen.

In the 1930s, when the British economist John Maynard Keynes wrote that in 2030 we would have to work no more than fifteen hours a week, he also asked himself if we would actually manage to do it—if we would be able to master the art of living without completely filling our days with work. What would happen if we suddenly had more time? In his book *In Praise of Idleness*, the philosopher Bertrand Russell writes:

> Suppose that, at a given moment, a certain number of people are engaged in the manufacture of pins. They make as many pins as the world needs, working (say) eight hours a day. Someone makes an invention by which the same number of men can make twice as many pins as before. But the world does not need twice as many pins: pins are already so cheap that hardly any more will be bought at a lower price. In a sensible world, everybody concerned in the manufacture of pins would take to working four hours instead of eight, and everything else would go on as before. But in the actual world this would be thought demoralizing. The men still work eight hours, there are too many pins, some employers go bankrupt, and half the men previously concerned in making pins are thrown out of work. There is, in the end, just as much leisure as on the other plan, but half the men are totally idle while half are still overworked. In this way, it is insured that the unavoidable leisure shall cause misery all round instead of being a universal source of happiness. Can anything more insane be imagined?

Sadly, Bertrand Russell's words are as pertinent today as they were in 1932.

"Bertrand Russell claims that idleness is *not* the root of all evil—but the root of scientific discovery and the creator of great art and music. Busy offices produce neither good science nor good art. I'm focused on us all having a better life and a better society, and that means we have to try working less. Growth is not the solution; the idea of growth is in fact a problem for all civilization," says Anders Fogh Jensen.

Both authors believe that to admit doing pseudowork is one of the biggest taboos of our time; it's something you cannot even reveal to your partner or best friends. There is so much shame involved. Your job should be important and have a purpose and, if necessary, continue beyond your working hours. In the

Protestant iron cage, the only thing that matters is work, not enjoyment or leisure—and yes, there are other iron cages too, ones that reward purposefulness more than nonpurposefulness.

"The fact is that many of us do something that only resembles work. It's the illusion of work. And it's not just the hollow bureaucracy and monitoring of our own (and other people's) work; we also have meetings that never reach any conclusion, and read thousands of pointless emails. We do things that are not important, while pretending that they are," Dennis Nørmark says.

Nørmark is happy to be self-critical on behalf of the right, where it is taken for granted that workers are motivated by a carrot-and-stick logic: punishment if we work too little, a carrot if we do a great job. That fact that we can be driven by inner motivation has never been considered. The ideology behind New Public Management, where capitalist ideas are applied to the public sector, is that everything should be measured and counted: every single procedure in a hospital, every consultation with a psychologist, every child who needs special education—it is all surrounded by a colossal amount of bureaucracy that measures and counts things that were once left to individual discretion and competence.

"Most health care professionals want to give the most meaningful help they can to the people who require it. It's our job, and our motivation for doing it comes from within. But if those professionals are overridden, this motivation can be ruined—and I find that many of us now operate with a more defensive rather than proactive form of motivation, which increasingly involves filling out the correct forms, diagnosis coding, and generally keeping the medical records free of corrections," the psychologist Sondre Risholm Liverød wrote frustratedly in a recent article, in response to the increase of bureaucracy in mental health care.

"We have far more bureaucrats than we have people doing a proper and important job. More and more of the public sector is now governed by bureaucrats who monitor other people's work—while at the same time, we have a shortage of nurses and teachers," says Dennis Nørmark.

Paradoxically, we now have more access to tools that free up our time.

"Because of artificial intelligence, we now have a chance of living a Greek vision of society. The ancient Greek philosophers didn't think working was a particularly good thing; they had servants for that. We now have computers. And that means we can spend our time on other things, like being creative, or being together," says Anders Fogh Jensen.

Jensen has now started practicing what he preaches. He manages his own time. The philosopher now works as a writer, a lecturer, and a college professor—and has a much more creative life than he previously did.

"I think, for many of us, our lives aren't quite what we'd imagined they would be. I used to buy a lot of books, because I anticipated reading them, but I never did. Now I actually do read them. I think you can get closer to your real self if you work less—and on things you find more meaningful," he says.

"But if we had more free time, don't you think people would just spend more of it sitting in front of the TV?" I ask.

"That's not what most people want. The majority won't want to watch Netflix several hours a day purely because they have more time for it. I think we would spend more time together instead. Right now, people don't have time to be with their loved ones. That's where it will start," says Jensen.

"I would say we were alienated, as Marx describes it—detached from the means of production and the purpose of what we're doing. This work logic is so ingrained in us we're unable to see it ourselves. Getting everyone to stop is difficult. People stop when they get sick—but it shouldn't have to go that far," he says.

On the liberal side, Dennis Nørmark believes that one of the driving forces in modern work culture is the pursuit of material goods, a pursuit he believes is of no benefit to employees.

"Working hard so you can consume even more is great for capitalism. But the happiness ratings are the same, whether a person earns a lot or if they simply earn enough," he says.

So how can you live differently in a society that's driven by ever-increasing consumption? If I'm to believe the people I see on Facebook and Instagram, it's all about going on expensive holidays to exotic locations, having several cars, making sure your children are constantly entertained and sent to football training and karate lessons, taking care of your dog and maintaining the cabin, and topping off your week with some Caribbean cooking from a recipe you saw on a dazzlingly colorful food blog. Not only that, but you need to have a beautiful home, go to the gym regularly, and wear clothes that perfectly express your taste and personality. All this quickly amounts to a life with very little free time, and almost no time for the noble art of daydreaming.

Kaja Gjedebo is one person who decided to "earn just enough." The full-time jewelry designer, who has had success both at home and abroad, lives in a little house with her family, and doesn't dream about having anything else.

"I don't want people to keep buying more and more, and I wouldn't want to sell my jewelry designs to a chain store, even if they were to ask. I want people to buy a piece of jewelry for life and really appreciate it. I don't want it to be sold everywhere and become a mass-market item. I'm interested in Charles and Ray Eames, the designer couple who tried to democratize good design and make it available to people other than the very rich. They wanted to make valuable and durable items for the ordinary person. That's what I want for my own design," she says.

We sit together at the dining table in Gjedebo's little studio, eating toast and drinking coffee. She has ground the coffee beans in a tiny, bright-yellow electric coffee grinder from the 1950s, which still works despite being over sixty years old. I notice that most of her things are well used and well loved after decades of being repaired and polished.

"Repairing something has a creative value of its own," she says.

"I wish we could spend more time nurturing our day-to-day creativity. Think about it: in the past—and I don't mean very long ago either—everyone had to be more creative on a daily basis. We had to repair clothes and furniture, build extensions on the cabin without really having any materials, and sew pieces of an old blanket together to make a coat. We're a species that loves challenges and tasks, and finding solutions that weren't there before. I genuinely think most people miss doing something with their hands, and the nice feeling you get from having mastered something purely physical. And yet people sit in front of the computer all day, and in front of the TV all night," she says.

"I've tried software that measures everything digitally for me, but I still need to feel the weight of the jewelry in my hands, and how it feels when I wear it. I trust my hand-logic. There's something especially satisfying about holding something and making something with your hands. It's a science in itself," she says.

We don't, of course, need billions of full-time artists. But there are more than enough clothes in the world that can be patched up or resewn—which is perfect, because the clothing industry is one of the world's most environmentally damaging industries. Working less means having more time to be creative every day, and being able to repair your clothes instead of buying new ones. This also helps reduce CO_2 emissions and gives you more time to spend with the people you love—all because you are doing something creative. Gjedebo has just learned to patch her clothes in a very beautiful and creative way: using embroidery and trimmings. When they're patched like this, it's like they become a nicer pair of trousers. It's almost reminiscent of the ancient Japanese art of *kintsugi,* where broken porcelain is repaired with gold, making it even more beautiful than before.

"It was a muscle that was far more developed in the past," she says, pouring more coffee.

"We can't all become artists, of course, and I think it's strange when people quit their jobs to pursue some tiny little talent. It makes it so either-or. Either you're an artist, or you're not an artist. But why not just fill your life with more everyday creativity?"

So what do we want? A life full of work and consumption, or something totally different? Let's remember for a moment that most of those who worked creatively in the past belonged to the upper or upper-middle class—people who had plenty of time to think about life and art. The Brontë sisters were the daughters of a priest and lived a secluded life in Yorkshire, with little else to do but read and write. John Logie Baird, his father also a priest, had lots of time to think up clever inventions in his tiny Scottish village. And Lewis Carroll, of course, who played with his ten siblings, was shielded by the moderate luxury of his parents' home at the vicarage in Daresbury. Edvard Munch came from the middle class in Kristiania, and Henrik Ibsen too, from a bourgeois home, before his father went bankrupt. A middle-class life in the nineteenth century allowed more than enough time to massage the DMN. These people could sit and read, write letters, go for walks, and attend the theater; they had home concerts and trips to the opera, friendly get-togethers, and leisurely dinners. Today, most people in Western society have access to all these things, nearly all the time. So why don't we do them more often?

"It's not about individuals—it's systemic, it's political," says Siw Aduvill, about her silent revolution: rest.

Nearly 70 percent of people would sooner have more leisure time than higher salaries and more stuff. Yet we continue to work, faster than ever, more connected than ever, more entertained than ever, with fewer opportunities for us to go into our own heads and spend time with ourselves and our friends.

"There are days when I think my role as a yoga instructor is to pick up the fallen fruit of a system that takes very poor care of people. When so many people choose to do yoga, it's not just because they want to be able to stand on their heads or do the splits after the age of fifty. Those of us who teach people to breathe, stretch, and rest are part of a wellness industry that is growing proportionally to how exhausted people are. Job streamlining and rising levels of perfectionism just means more business for us," Aduvill believes.

But it's a job Aduvill doesn't really want. She wants us to become less stressed and more sustainable people, as a society. The recipe for Aduvill's "parasympathetic revolution" (or perhaps we should name it the "giraffe revolution," after the poster I saw at school?) consists of long dinners, wine tasting with friends, cell phones switched off, afternoon naps, togetherness, singing, laughter, sex, and generally enjoying ourselves. But all of these things are drowned out by the sympathetic mode people so often find themselves in today, a stress response that behaves almost like an epidemic.

"There's scientific support for claiming that it's healthier to live in a culture that's trusting and caring—and where we have time for each other," says Aduvill.

"We can't be frightened into feeling more secure. Hatred, fear, and anger are counterproductive if the goal is better health, more interaction, and networking—and less isolation, worry, competition, anxiety, and depression," she says.

Now, you're probably starting to think that I'm drifting a little off point. This is a book about creativity, right? But if the research is correct, and the professional creatives I've met are being honest with me, then it seems there's little to be gained from being in a constant stress response. If your goal is to think new thoughts, you need to be confident enough to have them. If you want to daydream, it's no good feeling scared. That's why *parasympathicus* is the creative network: it's where you can daydream without interruption, try out new thoughts and ideas, gather inspiration and impressions. If you're creating something totally new, such as an invention, a discovery, a work of art, or a book, and you want to share it with other people, you are in *parasympathicus*, the place for connection, empathy, friendship, and community—the state of mind that allows you to have the most radical ideas. You can't convey anything to other people if you don't think they will listen; you can't cooperate well if you're in a stress response. Writing this book, for you to read, is an act of kindness—something I am sharing because I want to be your (slightly fictional) friend (and partly because writing makes me happy).

That's why I'm pretty sure that when Edvard Munch painted *The Scream*, he was paradoxically feeling neither loneliness nor anxiety—despite that being precisely what the picture expresses. Munch felt joy. That also explains why so many artists tolerate working under such poor conditions, with little or no guarantee of success, no pension scheme, or sick pay, or proper salary—things that anyone else with a job takes for granted. (If you are an author, you might get one or two lavish publishing parties with free alcohol to look forward to.) So the question is: Does the joy of doing something you think is deeply meaningful make up for the loss of material goods?

My own life in *sympathicus* eventually came to an end. I couldn't take it anymore. So I quit my job. I had never imagined it would be possible to give up like that, but that's what I had done. I was now unemployed. The sound of the door slamming behind me for the last time made me bite my lip. I tried not to cry, but there was an unmistakable rumble of fear and confusion in my gut, threatening to take over my whole body. It was hard to believe that I'd turned my back on all the benefits of having a job, and that I now had nothing but a keyboard to keep me financially afloat. I stood on the street for a second. Where should I go? What should I do? It was the middle of the day, the sun was shining, not that I noticed—the sunlight on this spring afternoon just looked gray. After walking aimlessly around the city searching for answers, I had to start heading home. As soon as I was inside my apartment, I slumped to the floor and cried.

I have devoted a great deal of my life to work. I take great pride in being a hardworking person. I started by delivering newspapers when I was twelve years old, then moved on to working in a newsdealer. I have washed floors and worked at a grocery store checkout and in a bookstore; I worked as a journalist, then in PR and publishing. I have led panels and events, I have started a festival, and I have written books in the evenings after a full day's work. In short, I have worked incredibly hard, and often through long, twelve-hour days, for very little financial gain. I worked while going through a breakup, while the sweat dripped off me due to a high fever, and while I had a concussion. I was working the day a good friend of mine committed suicide. I have checked my work email on Christmas Eve, on New Year's Eve, on vacation, and in the middle of the night. I have always thought that if I worked hard enough and long enough, it would somehow pay off—that there was a reward waiting at the end of the tunnel, that someone would cheer, someone would say, "You did it! You made it to the finish line!" I have waited for someone to give me a hug or a medal. I have been an atheist yet behaved like a good Protestant. I forgot about the good things in life and devoted myself to self-sacrifice. I have *not* enjoyed myself every moment. And then, after all that, I had to stop.

After months of not really seeing my family or being able to relax, thrown into a crisis mode where I was too scared to check my email (in case it was my boss telling me that I'd done something wrong), something happened. In that time, I hadn't allowed myself much in the way of food or sleep, and I'd lost fifteen pounds; I had black rings around my eyes and my hands constantly trembled. My clothes hung from my body, and my hair was starting to break off. Then my husband caught a cold, and for the first time since beginning her new kindergarten, I had to drop off my daughter. I was already feeling mildly irritated at being late for work, and when I stood in the kindergarten, explaining to the staff why I hadn't been there since she started, it became clear to me that I was in completely the wrong place in life. I was somewhere the Cheshire Cat would never appear, but the Queen of Hearts would scream at me around the clock.

My little four-year-old tugged on my skirt with her usual, insistent self-confidence.

"Mama, I made this for you," she said, taking a drawing from her closet and placing it in my hand. Somewhere in the chaos of blue scribble she had fearlessly covered the entire sheet of paper with, I could just about see the outline of my own unhappy face. I knew what I had to do. I felt weightless, just as Alice's fall down the rabbit hole into an unfamiliar world must have felt. I thought of Lewis Carroll, who after twenty-six years of teaching math at Oxford quit his job to write full-time, and how scary he must have found it. Freedom awaited. All I needed was the courage to grab it.

7 | The Walrus and the Carpenter

OR: I LEARN HOW TO LIVE IN THE FUTURE.

"Living backwards!"
Alice repeated in great astonishment.
"I never heard of such a thing!"
"–but there's one great advantage in it,
that one's memory works both ways."

IN THE SUMMER of 1816, Europe was struck by exceptionally bad weather. The previous year, the Indonesian volcano Mount Tambora had erupted, sending huge clouds of ash into the upper atmosphere, which led to heavy rain—and even partial darkness in the middle of the day, as far away as Europe. It was not the lazy Southern European summer that the renowned British poet Lord Byron and his entourage had anticipated. If we are to understand what the conditions for creativity are, we may as well start here, just over two hundred years ago.

As lightning flashed across the Swiss sky, and thunderclouds rolled in from the dramatic mountain peaks beyond, Byron's group took shelter in the luxury mansion Villa Diodati. While the storm raged on outside, boredom slowly set in, until Byron finally suggested they sit down and tell ghost stories in the light of the flickering candles. First they shared well-known stories, then they began making up their own. Byron had brought his doctor along, and the other travelers included the poet Percy Bysshe Shelley, his girlfriend Mary, and their son Willmouse. Mary's stepsister, Claire Clairmont, who was Byron's mistress and pregnant with his child at the time, also accompanied them. None of them were concerned with the institution of marriage—or, for that matter, any of society's other rules and norms. They were upper-class freethinkers, inspired by the radical thinker William Godwin.

Surprisingly, of the stories told that summer, it was the contribution from Shelley's eighteen-year-old girlfriend, Mary—who happened to be the daughter of William Godwin—that went on to become the most famous. It was during this summer, just over two hundred years ago, that she created one of the central mythical creatures of our time. The character she scared the others with has, more than any other literary character, come to portray the shadowy sides of human creativity. Mary had been inspired by their conversations about galvanic electricity (named after the Italian scientist Luigi Galvani, who performed an experiment that made a dead frog kick its legs) and the pros and cons of the emerging field of modern medicine. Mary Godwin, later Shelley, created the story of a scientist who pieces together an abominable monster from different corpses and brings it to life using electricity. The novel was to be called *Frankenstein*.

It is also quite plausible that she was inspired by Frankenstein Castle in Germany, which they had passed en route to Geneva, and which centuries earlier had been home to an alchemist. Alchemists dreamed of becoming masters over life and death. These pseudoscientists were also interested in Jewish folklore—such as the myth of the golem, which is all about creating life out of dead matter. In the nineteenth century, one such story circulated in Prague, about a man who had been created from clay by Rabbi Loew in the city's Jewish quarter. What these stories had in common was also what made Mary Shelley's novel seem so feasible—she took our basic fears about creativity getting out of control and put them into words.

But the story of the freethinkers who sat writing ghost stories in 1816 is about more than a foresighted woman and out-of-control science. Because while the poets Byron and Shelley may be remembered best for helping to form the myth of the creative genius—after living short lives marked by an anarchistic disregard for conventions—it was the women they left behind—and their ideas—that pointed to the future. Mary Shelley (who became a widow at twenty-five) was one of them, with her story about Frankenstein's monster. Lord Byron, for his part, left behind a daughter who would be of crucial importance to the development of modern computer technology: Ada.

While her father became the paragon of creative, romantic genius, Ada's mother wanted her daughter to use her energy on something more tangible—it's possible she thought Byron was unstable as a result of his extravagant lifestyle and love of poetry. Either way, Ada grew up without her famous father; her parents divorced when she was barely a year old, and she never had a relationship with him. Lord Byron died at just thirty-six years old, fighting for the Greeks during their struggle for independence against the Ottoman empire.

Ada Lovelace is now viewed as the world's first programmer, mainly due to her unique ability to creatively use math to see into the future. She had no access to computers, of course, since this was the first half of the nineteenth century; nevertheless, it seems that she was still able to predict some of the further consequences of modern mathematics and technology. And although her mother made great efforts to help Ada become a sensible mathematician, the girl seems to have been equipped with a powerful imagination—something we now know made her an even better mathematician.

In 1833, when Lovelace was seventeen years old, she met Charles Babbage, a twenty-five-year-old family friend. Babbage was a prominent mathematician and an associate of most of the celebrities at the time—and had already used up the staggering 17,000 pounds granted him by the British government. The money was to be used to build a so-called "Difference Engine," a machine that would be able to more economically calculate the astronomical and mathematical tables so important to the island nation's shipping and commerce. Despite all the funding, the machine never became a reality, and by the time he met Lovelace, Babbage was working on a new calculating machine, the "Analytical Engine," which he hoped would be able to perform all sorts of mathematical calculations. This time, like the Askeladden he was, he used an invention that Joseph Marie Jacquard had made to revolutionize the textile industry: punched cards. Essentially, storage devices, based on "zeros" (where there are *no* holes) and "ones" (where there *are* holes).

Ada had long shown a great interest in and understanding of mathematics, and her mother made sure she had private tuition from an early age. The young lady eventually married a lord and had several children. But she was always passionate about mathematics, and she wasn't exactly lacking in confidence when it came to her own talent.

"I do not believe that my father was (or even could have been) such a Poet as I shall be an analyst," she wrote to Babbage in 1843.

Babbage, at the time, had given her the task of translating an essay that Luigi Menabrea, the Italian engineer and later prime minister, had written about the Analytical Engine. Ada understood her good friend's invention far better than Menabrea—and it was in the translation that she made her most important contribution to the history of programming, because her translation notes were actually far more comprehensive than the essay itself, and far bolder.

In Lady Lovelace's Note A—the one that made her famous—she launches the radical idea that an analytical machine of this kind might be able to perform other types of calculations, not only mathematical ones. She suggested using a kind of symbolic language to make the machine do something other than mathematical operations, "whose mutual fundamental relations could be expressed by those of the abstract science of operations, and which should be also susceptible of adaptations to the action of the operating notation and mechanism of the engine."

This is where modern programming history begins, with this description of programming, almost one hundred years before the world saw the first computer. In the now-famous notes, Lovelace even implies that the machine would be able to compose music.

And she was right. Computers do make music. There are now algorithms for making generic music, for writing generic news items, and for writing generic literature.

"It is desirable to guard against the possibility of exaggerated ideas that might arise as to the powers of the Analytical Engine... The Analytical Engine has no pretensions whatever to originate anything. It can do whatever we know how to order it to perform," wrote Lady Lovelace in Note G of her translation, for the first time referring to the idea that a machine may one day have the ability to think for itself. The computer pioneer and mathematician Alan Turing later called this "Lady Lovelace's objection."

Alan Turing invented the so-called Turing test, which was designed to recognize the difference between machines and humans—and which no computer has ever passed. In addition to Lovelace, Turing is responsible for having the greatest impact on the development of modern computers: he created a modern version of Babbage's Analytical Engine—and under extreme pressure too, not terribly ideal circumstances for being creative. During World War II, the mathematician led an entire team on an assignment to crack the German coding machine "Enigma," and many believe it was here the war was won, at Bletchley Park, a beautiful house set within a walled garden (Oliver Sacks would have nodded approvingly). Perhaps we could say that the creative circumstances were ideal after all. Nevertheless, it was at this facility in Milton Keynes just outside London that Turing designed the first real computer—which is how the British managed to crack Enigma and read the Nazi regime's strategic military communications. As early as 1936, in the article "On Computable Numbers," Turing had described what he called a "universal Turing machine," a machine capable of making any calculation that could be described in an algorithm. It would be able to solve not just mathematical problems but any problem at all, just as Lady Lovelace had envisioned almost one hundred years earlier. Turing's code-breaking machine "Bombe," created for the British authorities in the service of peace, would have enormous consequences for civilization after the war. But we must first take a quick look at the work of a Persian mathematician who lived in ninth-century Baghdad.

Muhammad ibn Musa al-Khwarizmi is known as one of history's great mathematicians. The founder of algebra and one of the first to use the number zero, al-Khwarizmi translated Greek and Sanskrit manuscripts into Arabic during the Islamic Golden Age, and in doing so gave us some of the first systematized algorithms. An algorithm is really just an instrumental and step-by-step recipe for solving mathematical problems. With this type of symbolic language, one could analyze mathematics on a more logical and abstract level, and solve advanced equations. The word "algorithm" comes from a misinterpretation of al-Khwarizmi's name, which was incorrectly written as "Algoritmi" in Latin. And now al-Khwarizmi's name is linked to what many think of as a terrifying future, where artificial intelligence will take over human tasks and start thinking for itself—and might even become creative.

Algorithms can be easily used by a computer program, and modern computer programming has now made it possible to create algorithms that give computers learning tasks, such as the ability to recognize faces or animals by scanning an infinite number of examples. Since computers work incredibly fast, they can quickly acquire vast amounts of knowledge. Just as a child will combine information and experience, computers can be taught—but unlike a child, a computer will not suddenly come up with something surreal, like chocolate-milk-juice. Or could it be that we're about to face a new type of computer that's uncomfortably similar to ourselves?

"You can set a thousand monkeys, or a thousand computers, a writing task. And they'll all write mind-blowingly fast, but most of it will be worthless. Our problem isn't that there's not *enough* art being produced. We have thousands of finger-painting preschool kids making sure of that every day. The problem is recognizing the signs of genius—what's beautiful, or meaningful—and refining it. Machines aren't much use to us on that front," says Wilhelm Joys Andersen, one of Norway's most talented programmers.

Andersen has founded several companies, and for seven years worked as a developer and manager for the internet success story Opera. He turned down a job working for Google to instead develop Norway's online encyclopedia, and has several times been proclaimed one of the country's greatest leadership talents. Originally from Bergen, he is a distinguished man in his thirties—exquisitely polite and well dressed in a 1950s suit and tie. He was previously in Norway's left-wing activist group Rød Ungdom (Red Youth), and certainly doesn't buy into all the IT industry clichés.

During his programming career, Andersen has made himself redundant multiple times. That's right, he has programmed so well that he has succeeded in replacing himself over and over again. But he has risen from the ashes every time, since there have always been new projects waiting. The important thing to remember here is what computers cannot do—and that's something Andersen will always be the best at.

"It's something I can never put into a program. Wait here a minute," he says before leaving the room. I wait.

Then he returns, carrying a large box, which he throws on the table. The box contains a heavy bundle of A4 sheets. Had he said it was the collected works of Karl Ove Knausgård, I would have believed him. But it's not that. The papers are covered in densely written code.

"These 3,500 pages of code amount to four months' work, for three programmers. About a year's work altogether. Programming is a subject that looks very abstract and intellectualized. But it's like molding clay, actually. My Japanese in-laws have been producing ceramics for hundreds of years, and I recognize myself and my programming in their work. In much the same way, programmers have to feel their way forward until their work is the right shape. It's tactile, and sits in your hands. Many of the steps in this process can be automated, but the actual *feeling* of solving something correctly is something a computer cannot understand itself. Nor can it have the same overview that tells you if you've solved the right problem," he says.

His creative work, making his digital "ceramics," began when he was six years old and was playing around with a Copam 286, his father's PC, in the library at home. He found the source code for one of the games and adjusted both the high score and gravity settings, so that when the pixelated monkey threw bananas in the air, they never came down again.

"I had altered the laws of nature, in a microscopic universe. Like a tiny god. It was my first taste of programming. From the age of twelve, I was quite sure where I was heading," he says.

The code sitting on the table between us is for an automated real estate service, which in the space of two years has sold properties to the value of 430 million dollars—and is just one of several development projects that he has led. Now Andersen is the general manager of Minus, which has thirteen employees and an annual turnover in the millions. He is also the technology manager for the banking service Folio, which has twelve employees. Not only that, but Andersen has a leading role in the internet world; he is a member of the World Wide Web Consortium (W3C), which determines the standards for how the internet should work, a forum where seven hundred of the world's most powerful engineers and developers, people from Google, Apple, Microsoft, and Mozilla, make decisions that have major consequences for us ordinary people. There are, after all, over four billion internet users worldwide.

"In the last twenty years, the internet has gone from being a strange hobby for some of us, to a societal infrastructure. None of the big internet companies, or W3C, have taken into account what this really means," he says.

What's special about the internet is that it amplifies everything a thousandfold, leading to both amazing and unpleasant consequences. There are now tools for bringing people together, spreading ideas, and making work more efficient, tools that can make us more creative. But every good thing that's been made has also become a weapon in the hands of the world's pedophiles, neo-Nazis, misogynists, fascists, and unscrupulous capitalists. *Everything* is a thousand times more powerful.

"If you want technology that works, you have to listen to the technocrats. But Silicon Valley technocracy on its own is the fastest way to dystopia. What's needed here is a humanistic corrective," he says.

Andersen's job primarily involves facilitating creativity, and the working environment he created for his staff is entirely free of control. There isn't a time clock in sight. The staff could be anywhere in the world, and they usually are. While he and I are sitting there, only one of them comes in to work; otherwise the place is completely empty and quiet. It is work culture 2.0. In a business where every single employee can be responsible for a huge, multi-digit turnover, and where creativity is totally crucial, the rules about working hours are the least stringent.

"Why in the world should I control my employees? This isn't the assembly line from *Modern Times*, where everyone has to hammer in unison. Our work involves both collaboration and alone time, where you're in deep concentration. Most often, it's the latter. Leaning across your desk to ask a developer a quick question when they're deep in *the zone* costs an hour in lost productivity. It's like disturbing someone making a ceramic. That person will then have to wash the clay from their hands, take off their apron, and come right out of their thought flow. The clay will harden, and they'll have to start again."

If computers are our clay golems, it's no wonder we imagine them coming to life, gaining consciousness, taking over the world, and ridding it of humankind. We're now afraid that we have created a monster with greater and even more incredible powers than us.

"Imagination is one of the most unique human functions, so the question is whether computers can imitate just that," says Per Kristian Bjørkeng, who has written a book about artificial intelligence.

"It's far more relevant to talk about pattern recognition. But that alone has brought computers a very long way, and it's surprising how far they seem to have come. You can now type in the first sentences of a fairy tale, and a fairy tale will come out. They can create a photo of something that doesn't exist. They can write poems and speak with believable voices. Right now, we don't know where it will end; there's no wall to be seen," says Bjørkeng, who is a technology writer for one of Norway's most-read newspapers and follows the field closely.

"If a machine does something quite different from what it was trained for, then something is wrong. That's where the limit lies.

Still, you can train machines to do things that most people won't expect. You get the creativity you ask for, but nothing more," says Bjørkeng, echoing the words of Ada Lovelace.

A computer won't suddenly come up with an idea for a helicopter, like Leonardo da Vinci. If we ask one to draw a picture of a cat, for example, it won't be able to have surprising, spontaneous ideas about a bicycle. Just as it was in the mid-1800s, a computer will be able to do only what we command it to do. The difference now is that we're asking computers not just to calculate "54 × 87," but to learn how to play complex board games like the Chinese game Go. Go is a very complicated strategy game played with black and white stones on a large board, which one of the opponents has to conquer. It's not unlike chess, except that it involves far more strategies and possible endgames. When a computer managed to beat one of the world's foremost Go masters, it was an even bigger watershed than when the supercomputer Deep Blue beat Garry Kasparov at chess in 1996, an event that, at the time, sent shock waves around world. But it took another twenty years for artificial intelligence to master Go, a far more advanced game, when Google's AlphaGo beat the European champion and then the top-ranked South Korean player, Lee Sedol. Finally, in May 2017, AlphaGo beat the world's best Go player, Ke Jie. The task had been accomplished: a machine had learned a game requiring intuition, experience, strategic knowledge, and an overview of an infinite number of possible smart moves. In addition, AlphaGo has expanded the game and changed the tactics considered to be good moves. What's so radical about Go is that it is a cumulative game, becoming more and more advanced as it progresses—the opposite of chess, which gets easier as pieces are knocked out and the game reaches the end. Go is a very creative game.

"Artificial intelligence can't put totally different ideas together and create new ones—yet. But what exactly is creative work? In the art world, Damien Hirst has assistants, and they do a creative job too. Having a computer make something for you isn't fundamentally different. You make the big decisions and point out the direction. Your artificially intelligent assistant fills in the details, but has a significant amount of creative freedom and can deliver a high level of craftsmanship," says Bjørkeng.

It is tempting to think that what's most creative about artificial intelligence is that it could free up a lot of time for us humans, allowing us to be more creative than before.

"There won't be less people working creatively. But the routine jobs will be replaced and automated," Bjørkeng believes.

The least creative jobs, the ones that are easiest to get a computer to do, will eventually no longer be jobs for people. That should allow us more time to work on other things. So the question is: Will creativity increase? And will productivity increase? Will we have a supercharged work life?

"With machines providing more and better help, we'll be able to generate more. As a journalist, I can ask a computer to do research for me, and it will go through hundreds of thousands of articles, more than I ever could. We can use this extra time to increase our leisure time, or to increase our income. It is a huge, collective issue, something the trade unions have to work on. Because we can ask for more time. We can be like the Greeks, who spent their days thinking beneath the arcades, or we can spend more time with our families. We have to create a society around people, not machines. What's happening now is a real opportunity to make big changes and prioritize our lives differently. But it's also quite possible that the big changes on the way will be more destructive than constructive, if we are not careful," he says.

What we are all most afraid of is that *our* jobs will disappear, and I do feel a little anxious right now at how closely knit my laptop and I have become. Maybe *I'm* redundant? But can machines have a *perspective*? Can they have a *voice*?

"It may be possible for genre fiction to be partially automated—I mean, the parts that can be written by a computer—and then have a human transcribe it. Humans still have to be the ones directing and making the big decisions. In practice, a machine can only create new variations of something it has seen many examples of before—it simply cannot create something that is both good and genuinely new. Human experiences, things that are unique and rare, are things a computer will not be able to re-create, and if it did, it would be of no interest to us. I wouldn't have had the same experience reading Knausgård if I thought a machine had written it," says Bjørkeng.

The author Bår Stenvik is not as optimistic as Bjørkeng. Stenvik quotes from Peter Wessel Zapffe's 1933 essay "The Last Messiah," where the philosopher writes about how man ruins everything for himself:

> Life had overshot its target, blowing itself apart. A species had been armed too heavily—by spirit made almighty without, but equally a menace to its own well-being. Its weapon was like a sword without hilt or plate, a two-edged blade cleaving everything; but he who is to wield it must grasp the blade and turn the one edge toward himself.

In Zapffe's interpretation, man will cause harm to himself with each step forward he makes.

"Artificial intelligence will be able to harm us too—not just give us more free time or make us more creative," says Stenvik, who writes books about AI and has previously written one about computer games.

Stenvik is interested in power structures. He doesn't believe that artificial intelligence will give us more time to be creative. It will simply become more competitive for those who want to.

"All the small jobs, the bread-and-butter jobs for people in creative professions, will disappear. AI can now make film music and write news stories, create photos and drawings. Very few would say that it's terribly creative—it's not a threat to people with great artistic ambitions—but for those who create stock photos and illustrations, or compose film music, or make mass-produced art for IKEA, it will be the end of the road. Then you have all the jobs in the production and distribution of art—proofreaders, people who write scores and master records. Those jobs will be gone," says Stenvik.

Most of all, Stenvik worries about how we are giving away power and information to people who don't have our best interests at heart. He does not think society will improve or become more creative as a result.

"When Google gets its cheap laptops into our schools, it will mean the entire next generation will be trained using Google services—and at the same time, Google will get all the information about how children learn. So Google will be sitting on all the expertise about how today's children absorb knowledge, and our schools will then have to buy it back from them. We are providing Silicon Valley billionaires with all the information they need about our own health and welfare, but this information isn't being used for the common good. Eventually, we'll become like the people of Haiti—who stopped growing their own rice because they were outcompeted by imported rice from the US, and then didn't have any of their own resources to fall back on when prices tripled later," he says.

There are many forces exacerbating the decline of different types of creativity—one being the disappearance of odd jobs, another being how services like Netflix and Spotify sap our attention and make us less open to new forms of expression. In many ways, these companies perform the same service that we do to ourselves when our brains are "pruned"—the branches are cut and our opportunities become fewer. So on a cultural level, our ability to think unconventional thoughts or understand unconventional creativity is getting worse.

"Only 10 percent of the music on Spotify accounts for 99 percent of the streaming. Its algorithms streamline and direct our attention, and this will lead to less room for human creativity, less diversity, and fewer genuinely new and strange ways of expressing ourselves. There will be less room for really creative innovations—unless the algorithms are trained to find them," he says.

Stenvik compares us to a group of nuclear physicists who have hired a magician to perform at the annual nuclear physicists' conference. We may be able to understand the most abstract and mysterious laws in the universe, but we're still allowing ourselves to be fooled by the magician's sleight of hand.

"We have predisposed cognitive weaknesses, and companies like YouTube know exactly what's required to tempt us into clicking on the next video link. Even so, I'm optimistic. Books still have a value, something to do with timing and perspective. It's a relationship that builds over time, a proper relationship. There's a big difference between googling something and reading a book. Right now, podcasts are very popular, which I think is a kind of backlash to all the fast online information we consume. Because the act of listening forces you into a slow mode; it creates something more reminiscent of an intimate, human friendship; the voice is totally inside your ear," says Stenvik.

Siri Hustvedt believes that the idea of a creative and independently thinking artificial intelligence is a primal and deeply human fantasy.

"Masculine birth is an ancient idea—think of Zeus giving birth from his head, Pygmalion, or Plato's pregnant male philosopher. Men have long dreamed of giving birth without women, without female reproductive organs," she says.

When we talk about all the bursts of electronic and creative innovation we believe are on the way, it's important to remember that we are the ones holding the compass, who can decide what direction we take. It is we who are dreaming the future. The reason our memories are so unreliable and elastic, and not at all like an audio recorder or film camera, is probably because we use memory as a building block to make stories about the future. What we dream of happening is usually a version of something we have already experienced. We cannot see all the other unspoken possibilities fermenting in the present. So we could say that science fiction is perhaps the oldest literary form, because that's how our brain works: we interpret present trends so that we can predict what will happen in the future.

At first, we did it to avoid being attacked by lions, whatever we could do to predict where they might show up next time; we planned how to arm ourselves with weapons, and we made workshops to make new weapons if we lost the first ones. Today, our fantasies stretch much further into the future, and are far bigger. For millennia, people dreamed of traveling to the moon, something we achieved in 1969. Now we dream of going to Mars.

"These fantasies have fueled both AI and robotics research. Many scientists were inspired by *Star Wars* and *2001*, for example. But I don't believe conscious machines are possible. Society will become increasingly automated and ruled by the algorithm. But automation does not mean being awake, feeling, and aware," says Siri Hustvedt.

The question is whether the brain's myriad associations and strange connections can actually be copied by a computer, and whether we really want that. In the human brain's daydream mode, there are all sorts of impulses that can lead to new ideas and discoveries—but is this something a computer can replicate? What's happening now with artificial intelligence is perhaps a watershed in the history of humankind. Or maybe the watershed came with Babbage's machine, or when Ada Lovelace finished her essay, or with Alan Turing's article in 1936, or when his code-breaking machine Bombe was ready to decode Enigma in 1940. Perhaps the watershed occurred when the first personal computer went on sale in 1975, or in 1990 when the World Wide Web was invented, or when the first iPhone was launched in 2007. But maybe the watershed has been happening constantly, because artificial intelligence is the result of the bold thinking and brilliant and foresighted ideas of so many creative people—brilliant ideas that, paradoxically, could take something from us, and bring other forms of creativity to an end.

"Of course, I want my drawings to change the world, but I don't quite know how. I just don't want it to carry on like it is now," says Siri Dokken.

One of Norway's few remaining satirical cartoonists working in the national press, Siri Dokken has worked for the newspaper *Dagsavisen* as its political illustrator for twenty-five years, and is also a professor at the Oslo National Academy of the Arts. Dokken is known for her elegant and pointed sarcasm disguised by beautiful line work. She aims to make us laugh at the present while looking into the near future. Yet her job is under threat from several factors. The reason there are fewer newspaper cartoonists is because they are a luxury few editors indulge in; editors can buy illustrations from an image bank or use photography, or not use anything at all. In the future, they might just feed an algorithm into a computer and use whatever generic result comes out.

In the past, Dokken's cartoons would be sprinkled with intellectual references and obscure puns. Now the world is burning in a different way, with satirical cartoonists being fired or under attack from extremists. When Norwegian and Danish flags burned in the Arab world in response to the Muhammad cartoons and many cartoonists had to be given twenty-four-hour security, Dokken's already vulnerable job suddenly became far more dangerous—even more so when thirteen people were murdered at the offices of the French satirical magazine *Charlie Hebdo*.

"I'm getting older, and not as vain anymore. I don't think I have time to be as convoluted, and I don't have the luxury of being sophisticated. We're in a climate crisis, and democracy is being undermined. It's more serious now; there's more at stake, although you can't say it. As a satirist, you can't show people your hand—it would be like Batman without a costume; you wouldn't get the same element of surprise. But now I'm trying to appeal to a community that goes beyond the political trenches. The personal has become very political, and vice versa," she says.

Dokken works at all times to connect with shared reference points, the cultural soup we're all swimming in, our common DMN, the spirit of the times. A newspaper cartoonist has to appeal to something people recognize.

Among the global driftwood, there are common references that can make us laugh, if they are connected in a new and unexpected way.

"I see the irony in this: I come from the irony generation, but something has to change; we can't carry on like we did before. If we don't do it ourselves now, the climate crisis may well decide it for us. We really need to talk seriously. There are masses of people searching for a community, yet there are many unconstructive communities on offer. I'm just a humorist who wants to remind you that we have common interests, that many of us have children, that the climate crisis is catching up, and that we're all afraid of dying," says Dokken.

Dokken will read the news piece she has to illustrate in the evening, and at night she'll allow her dreamworld to knead it into shape. Then at some point in the morning, between sleeping and waking, her drawing will appear. Snoozing has become her secret weapon.

"You can't make water flow in a specific direction, but if you let it flow where it wants to, things happen. It happens if you let go, as opposed to if you try hard. When I have a good idea, I can almost hear it—like a clear tone, it rings, and it's constant," she says.

In 2019, Dokken won bronze at the prestigious World Humor Awards for a drawing based on a simple idea, which resonated with our connected era. She had responded to a project called "Fifty Years After the Moon Landing" by drawing a little girl staring up at a full moon, which none of the adults can see because they are so preoccupied with their cell phones.

What will it be like after fifty more years? It's not hard to imagine a future where technology has become more important to us than both nature and human communities. A future where technology has become an integral part of us.

"We're a part of technology already," says the author Thure Erik Lund. "We're interwoven with it. I think and create with my computer."

It is therefore easy for us to imagine a similar future.

Lund believes that our ideas about the future are so strong they have almost taken control of the present.

"Our ideas and notions about the future affect virtually all technological activity. It's something that has also resulted in a significant expansion of the now, so much so that many of us are losing the intense feeling of a real now," he writes in the essay collection *Romutvidelser* (Room expansions). He believes that much of this was decided for us forty years ago.

"We're going in this direction now, we chose it back in the 1980s, and we can't turn around, not now. It's like when we chose between direct current and alternating current—we're set on that path now, where the possibilities have been set and narrowed down," he says.

In the late 1800s, there was a moment when an alternative view of technology and society could have been established, when William Morris and the Arts and Crafts movement launched the idea of an entirely different form of society—one that was in touch with nature and the means of production. Now, we are interwoven with a technology that has made the entire past available to us, and the dream of a high-tech future a powerful commercial and political force.

"In fact, we are all more or less unwitting suppliers of raw data. The endeavors of the vast majority of technology-dependent people have therefore ended up existing in one huge, transgressive structure of meaning, which is in turn aimed at helping to develop and improve technological systems," Lund writes.

Since the 1800s, technology has gradually become all-encompassing, a huge organism that is turning humanity into an interconnected "we," an eternally communicating commercial structure.

At the same time, we are entering a future that is more diverse than most of us can really visualize. What we often imagine is an extrapolation of what we fear in the present. We live in the past, and our memories of the past hurl themselves into the present and determine how we envision the future. The futures we imagine are full of the fears and hopes of the present—and the nightmares, dreams, and longings we carry from the past.

It's time to return to the full moon that shone over the villa near Geneva just over two hundred years ago. In 1816, the seasons did not behave as they should—it became known as the "Year Without a Summer." This was a taste of how things might become, perhaps, should we fail to achieve the two-degree target set by the Paris Agreement: seasons that are not as they were, extreme weather, and failed harvests. It was during this dark summer that Mary Shelley came up with her story about the scientist Victor Frankenstein, a story that was unearthly and frightening.

In Shelley's novel—now considered one of the world's first science fiction stories—Frankenstein creates a monster out of severed body parts and brings it to life using electricity, without fully anticipating what he is unleashing: a lovelorn creature that cannot find its place in the world, quite different to the artificial intelligence we've created from algorithms and internet data. Frankenstein's monster is lonely and needy, like a child waking up to a world he doesn't understand. He tries to find out what it means to be human, and reads Goethe in an attempt to understand. Robots are portrayed similarly on film: they wonder if they are not really human, they fall in love, they fear death. It's a very long way from reality. We may have created artificial intelligence, but we have not created life—not life like ours; not an irrational, associative, emotionally driven life that can spontaneously come up with ideas for cathedrals and space travel as a result of getting lost, or from resting, or from being quiet. It's important to remember that science fiction stories are not real predictions of the future. Like most fables about the future, they are instead a distorted reflection of the past. Yes—the past. So what past was Mary Shelley projecting into the future?

While recounting her ghost story, Mary probably had her little son Willmouse on her lap or in her arms; he was still just a baby. The previous year, she had given birth to a small, premature baby girl who lived for only two weeks. Mary had found her lying motionless in her bed on the morning of March 6, 1815, and her grief would be echoed later by the deaths of several more children. Of the four children she had with Percy, only one survived. And I have no evidence of this so I can only speculate, but as Mary sat cradling Willmouse's warm little body in the house in Switzerland, with the storm raging outside and the warm rain lashing the windows, she must have thought about the death of her baby girl, which had thrown her into a depression the previous year. She must have felt the imprint of the girl's small, limp body in her arms, the girl who appeared to be sleeping peacefully when her mother found her. Had Mary Shelley retained some power over life and death, some insight into the miracles of electricity and medicine, she may well have wanted to bring the dead girl back to life. Who wouldn't have wanted that? She must have dreamed about returning life to her dead child's body, a frightening creature made of body parts, unknown to the human world, longing for love and community. More than a story about the future, *Frankenstein* is perhaps really about the dead child, about unfathomable love and inconsolable grief.

8 | I Find Alice

OR: WEAVING IS BELIEVING.

"Everything's got a moral,
if only you can find it."

"YOU HAVE TO learn how to knit," says Vera, typically assertive. Never one to give up, and full of loving enthusiasm, she now wants me to do something that *she* loves doing. I've known Vera for ten years, and, in that time, I've repeatedly said how much I *hate* knitting. No, I *detest* it. But I also know that there's a climate crisis looming, so we actually should be making our own clothes instead of buying them, and besides, I realize that a little small-scale creativity is good for me. I also find myself going into daydream mode a lot quicker when I knit. And the other reason, perhaps the most important one, is that I want to make her happy. So for Christmas, I buy myself a knitting book, full of photos of gorgeous sweaters. I feel motivated. Vera shows me how to knit by following the patterns.

"If you want to be a good knitter, the most important thing you have to learn is to undo your stitches," she says, forcing me to undo a row I've messed up. I do a lot of undoing. But after a while I find my rhythm. With my fingers going automatically, my mind wanders freely in daydream mode, while a blue-green sweater based on a complicated pattern slowly emerges below. It's both useful and quite enjoyable.

"I've coped with every life crisis I've ever had by doing handicrafts," Vera tells me. She attends ceramics courses and has also learned to weave. So "Weaving is believing" is a slogan Vera uses frequently. I normally laugh at it, tease her politely, and call her a hippie.

But Vera is an author first and foremost, and, as with knitting, it was she who persuaded me to become one too. For years, I was so hamstrung by my inner critic that writing a book felt impossible. When I was twenty, I already had a publisher but, when it came to the actual implementation, paralysis set in. My inner critic simply grew into a giant. But all that changed when I met Vera. Vera spurred me on through my first novel, read every draft, and never gave up on me; her belief that I could do it never wavered. The critic on my shoulder may have been big and strong, but Vera was the angel on my other shoulder—and she was much, much stronger. By undoing everything and starting again, repeatedly, I finally managed to knit together everything I needed to make a nice and colorful book.

Vera had by then already written two novels for young adults and several nonfiction books for children and adults, and had a column called Ask Vera in the newspaper *Aftenposten*'s children's edition. She is also a much-loved children's TV personality, and whenever we are out together, we often meet people in their twenties who grew up watching Vera on the screen. Communicating to children, and taking them seriously, has been Vera's lifelong passion, and there seems to be no limit to her creativity. She is one of the kindest, wisest, funniest, and most vibrant people I have ever met, and the more I learn about the nature of creativity, the more I realize that she is creativity *personified*: she is rarely stressed. She is curious, open, unafraid of making mistakes; she thinks outside the box; she is totally disinterested in status and positions of power and glory, but very interested in all the wonderful things that happen when you explore the world. What matters to her is celebrating friendship and love and good food and flowers, bright colors, and fun stories. She is a *parasympathetic goddess.* I feel very lucky to have her as a friend.

So to make Vera happy, I'm learning how to knit. And because knitting clearly helps in a crisis, I decide to take it with me to the hospital, to sit beside Vera's bed in her glaringly white room, our knitting needles clinking and plinking under the fluorescent lights. Vera is knitting more than ever this winter, due to a life crisis requiring large doses of needlework and painkillers. Her cancer has returned.

It has slowly dawned on me while working on this book: something happens to our creativity when we are faced with death. And I'm not quite sure what it is, but I know who I need to talk to. Simon Critchley is a philosophy professor at the New School for Social Research in New York; he has written extensively about death, and is one of the founders of the International Necronautical Society. When the INS was established in 1999, the society members wrote a manifesto declaring that death is a place to be examined and inhabited—and that death is crucial in order to understand beauty.

"Death can be a powerful driving force in a creative process," he says.

Critchley has written books about suicide, the death of his father, Greek tragedies, the author Samuel Beckett, and the philosopher Emmanuel Levinas. Humans are unique in that we can acknowledge our own death and someone else's—and also have the ability to reflect on it. And this certainty of our own death can fill us with an urge to tell our story, before it's too late.

"Of course, the most obvious thing is that death is a kind of deadline, the final one, so it changes your perception of time. But someone close to you dying can start something in you; something happens. During the period my father was dying, I was working on Beckett, his trilogy about people dying, and it grew into a lecture, then a book," says Critchley.

Because how do we endure death if neither religion nor philosophy can help us? In a world where God provides only a pale shadow of the meaning to life, is it possible to find any answers?

"So, under the nihilistic conditions of modernity, the question of the meaning of life becomes a matter of finding a meaning to human finitude. In this way, we rejoin Cicero's question, restated by Montaigne, 'That to Philosophic is to Learne How to Die.' Our difference with antiquity, for good or ill, is that there is little sense of philosophy as a calmative or consoling influence that prepares the individual stoically for his passage on to either nothingness or eternal bliss," writes Critchley in his book *Very Little... Almost Nothing: Death, Philosophy, Literature*, which came out following the death of his father.

But we're not all similarly affected by death; every death is unique and sets different things in motion. When Critchley lost his mother almost twenty years later, it was completely different from when he lost his father. Both events, however, led to him writing a book.

"When my mother died, it was a whole other experience. While my father resembled one of Beckett's silent male characters, my mother was like one of the women. She talked so much, it was like a never-ending story. I couldn't believe she was dead, even after she was," says Critchley.

Time behaved differently immediately after her death: it became infinitely slow. Each second became a painfully long wait until the next. The philosopher, who had returned to England to be with his dying mother, flew back home to New York, and after only a few days was told that David Bowie had died.

"I wrote a book on Bowie, and put all my grief for my mother in that book, on top of the grief I felt for Bowie," he says. When Bowie died—following a fifty-year-long career in music—it was two days after he had released his final album *Blackstar*, now considered to be one of his best and most experimental albums. "Bowie turned death into art and art into death," wrote Critchley. Faced with death, Bowie managed to make a fantastic record, while continuously hiding the fact that he had terminal cancer. And in the end, the album's launch actually coincided with his death; one of the songs is called "Lazarus," after the biblical figure who rises from the dead.

• • • •

IN THIS BOOK, there's been a lot about how daydreaming, boredom, wakeful rest, and curiosity are important for creativity—so it might seem strange that death is too. Death would seem to be the antithesis of all these things. But with death comes a sense of necessity, where you have nothing more to lose because you have lost everything. This is precisely what the journalist and author Marte Spurkland experienced.

"It took a long time for me to realize that that was what I was doing. That the book was really about that," she tells me.

Spurkland recently published a book, *Pappas runer* (Dad's runes), that demanded all the energy she could muster. Writing a book about her father's work as a rune researcher was an intense experience.

"I've heard all the clichés, of course. I've interviewed plenty of writers as a journalist: 'The characters took over,' they'd often say, although I never really believed it," she says. "But when I worked on this book, I understood what they meant. I was totally engrossed. I could hear my father's voice while I was writing; it was like he'd come to life. It hadn't occurred to me that this book project would bring him so much closer to me."

Spurkland's book tells the story of her father, Terje, a runology professor—from the time of his diagnosis with an incurable brain tumor until it took his life approximately one year later. She traces his journey toward death, while also trying to learn as much as possible about his field of work.

"I wanted to strip the disease of its power and cruelty. My dad shouldn't be remembered for a disease. I wanted everyone to see his life's work, his almost fifty years of rune research," she says. When her father died at Christmas in 2018, Spurkland sat down to write, driven by a force that felt almost greater than herself.

"It shouldn't have been possible to do what I've done—to do so much work in one year. I've got two young children, a full-time job—and on top of that, I had this book project. But I was driven by something that made it impossible to stop," she says.

Spurkland had her father's notes for a book he was writing about runology—his life's work that was intended for the general public—and her own notes from the conversations they'd had about it. Runology was her father's great passion, but it was a field she had never properly studied. She learned how a rune stone would often be erected at the same time as a bridge, in order to help the dead cross into the afterlife. Its inscription told the world, and generations to come, about those who had passed away.

"It wasn't until the book was almost finished that I understood what it all meant. I had the story about Nordic runes, stretching from the third to the fourteenth century, and the story about my father dying of cancer. Then I realized: all runic inscriptions on stone were carved in memory of someone who died," she says.

Marte Spurkland's book is more than 432 pages of white paper, covered in black letters. It is a memorial stone. "Rune stones were memorials; so too is this book—I wanted to rage, rage against the dying of the light."

Writing a book about someone you've lost can feel like you're writing that person back to life, like you're engraving something onto paper, forever—a literary rune: "Marte raised this stone for Terje."

"I wanted the book to give my father eternal life, to really get to the core of who he was and keep him there. And it felt so compellingly necessary that it justified all the time I spent away from my children," says Spurkland.

Writing a book—any book—is like making a time capsule and sending it into the future, hoping that someone will find it, hoping that it will overcome time itself. Not only is it a profound act of friendliness, it is an almost desperate one, one that is full of hope. The prospect of death can inspire creativity, and the things we create are rune stones and bridges, memorials and gateways into our own experiences.

The psychologist Carl Jung was interested in symbols and myths. He wrote about a shaman—he who encounters a bear, loses an eye, and comes back a changed person. He has looked into the eyes of death and experienced the fringes of existence. He is twice-born, born again. Some people change after facing death; it can make them into more rounded, beautiful people who return with, and talk about, their experience.

"Some people believe in the gods, or fate, or whatever you want to call it; it opens a door for them—and they go where the rest of us don't dare," says psychologist Peder Kjøs, who thinks artists are similar to shamans.

When someone you love dies, it becomes extremely important to tell that person's stories—the ones they never had time to share themselves—to show the world who they were. The death of a loved one can bring about entirely new thoughts and feelings, and in the new landscape it produces, it can be difficult to know where you are. You get lost. Anything can happen.

Grief can even trigger the complete opposite reaction: laughter.

The comedian Else Kåss Furuseth's apartment is furnished with ceiling rosettes, well-kept parquet floors, and strange trinkets, awards, posters, crystal glass, and peculiar lamps in a riot of bright colors. After failing to find any matches to light the huge candelabra in front of us, Furuseth politely slides a porcelain corncob full of chocolates across the table to me. Had I not known any better, I would have assumed this was the Mad Hatter's tea party. It feels like visiting a fashionable Oslo home in a parallel Day-Glo universe; I almost expect the Dormouse to poke its head out of the teapot. The other thing I'm struck by is how many of the extremely fragile items in the room would be very easy to break, if you were to invite eight hundred people around, which she did.

In these blazingly colorful surroundings, it's hard to imagine the dark backdrop that made Furuseth a national treasure in Norway: both her mother and her brother killed themselves—and she has since become an outspoken and courageous advocate for openness around mental illness and suicide.

"I had an open house on World Suicide Prevention Day, because I truly believe, hand on my heart, that if more people crowded into small spaces more often, fewer people would feel lonely. And yes, I know you can't really have a massive house party every day, but I wholeheartedly believe that community makes people feel better. And having eight hundred people in here, in this apartment, is *almost* impossible," she says.

I can see it all so vividly: the porcelain smashing to the floor, the candles nearly setting fire to someone's hair, the coffee stains on the carpet, and the mud and dirt being ground into the fancy wooden floor by sixteen hundred feet. It was totally crazy. A crazy idea. Impossible, which, for Else Kåss Furuseth, means good.

"I like finding solutions for things that are really difficult. It has to be an impossible project that can be solved by a really idiotic, but simple, idea. Besides, I think the world is sad enough already, so I like to play around with serious messages," she says. "Anyway, it's laughter that keeps me going. I also love having to sacrifice something of my own, that it's not for free."

The reason we're sitting here now, eating chocolates from a porcelain corncob, is grief. One of the most important ideas Furuseth ever had came at a time when her world was profoundly dark. It was the day after her brother's suicide, and she had been looking through his weekly schedule when she noticed that he had been planning to watch the nine-hour-long Holocaust documentary *Shoah* that week, but never did.

"At the time, I said to my father, 'Had I known I was going to watch that movie, I would have killed myself too.' And we laughed together. Then my father said, 'But you must never repeat that to anyone, because nobody except us will understand.' So I wanted to find out if he was actually right. I decided to make a show about suicide, one that touched on all the taboos. Is it possible to laugh at something that's almost impossible to talk about? Not because there's anything funny about suicide, but because it's so incomprehensibly sad. And when something is that sad, I'm, in a way, extra happy that now and then something is still funny. I'm not out to hurt anyone; my dream is to make the audience genuinely laugh—not in a bittersweet or compassionate way, but because we're still alive," she says.

That doesn't mean she divulges *everything* in her shows *Condolences* and *Congratulations*.

"Suicide is still much sadder than it appears when I talk about it onstage; I hold quite a lot back. But my aim is simple: to find some joy for myself in saying what everyone already knows, but no one ever says. We have to be able to endure hearing that life is terrible, because maybe then more of us will be able to endure living it," she says.

It could be that William Shakespeare was processing his own grief through the protagonist of his play *Hamlet*, which he wrote several years after losing his eleven-year-old son, Hamnet.

"To be, or not to be—that is the question: Whether 'tis nobler in the mind to suffer the slings and arrows of outrageous fortune, or to take arms against a sea of troubles, and by opposing end them?" sighed the Danish prince from the stage of the Globe. But perhaps the question was one Shakespeare was asking of himself. Because how is it possible to endure the death of those we love?

Siri Hustvedt began investigating brain science more intensively after delivering a remembrance speech about her father at the university where he had previously taught. While standing at the lectern, she experienced severe tremors from her neck down.

"Confident and armed with index cards, I looked out at the fifty or so friends and colleagues of my father's who had gathered around the memorial Norway spruce, launched into my first sentence, and began to shudder violently from the neck down," she writes in the book *The Shaking Woman or A History of My Nerves*. Her father's death and the illness that followed changed her life in peculiar ways.

So the death of someone you love can trigger something new. We change, and there's no formula for this change—which is why sickness, death, grief, and heartache make us feel lost—and feeling lost is the start of a creative process. Now, this isn't meant to sound like creative advice or a tip of any kind. "Make sure that someone you love dies" is certainly not what I'm saying. But there does seem to be a connection.

"I tend to be drawn to what's missing," Haruki Murakami says of his writing. "You come across something missing, suddenly, when you don't expect it—like, you're walking across a field and fall into a dried-up well. I don't know why, but I'm attracted to that sort of situation. Something that should be there isn't, someone who should be there isn't. And that's when the story begins."

We humans are all the same; when faced by a huge loss we have a huge need for meaning. But it can be even more pervasive than that. When the neurologist Alice Weaver Flaherty experienced the death of her newborn twins, she was seized by hypergraphia—a manic urge to write—and it was so powerful that it kept her awake at night.

"Suddenly, as if someone had thrown a switch, I was wildly agitated, full of ideas, all of them pressing to be written down. The world was flooded with meaning. I believed I had unique access to the secrets of the Kingdom of Sorrow, about which I had an obligation to enlighten my—very tolerant—friends and colleagues through essays and letters," she writes. As a neurologist, she wanted to understand more about what had happened to her, and found that what she experienced was just an extreme variant of something everyone experiences when they suffer such profound grief.

A major traumatic life event can affect the entire limbic system—an area of the brain that controls hormones and reproduction, and which is also linked to the hippocampus, which coordinates memory. It is here that powerful emotions are linked to our memory. Grief of any kind can penetrate deep into the limbic system, causing you to rewrite your entire life story. These experiences will be the material of your most distinct memories. It's no coincidence that so many pop songs are about unrequited love or heartbreak.

"Love of another, especially if it is unrequited, is a threat to self-esteem—and self-esteem is something we think words can fix. Writing and talking to whoever will listen to us rises especially from the anguish we feel when the beloved is absent. The words we write, the stories we tell about her to ourselves, serve us, Pygmalion-like, as our artificial Galatea until she returns. In bereavement, the need to tell these stories can become unbearable," writes Flaherty.

The limbic system's most direct interaction with language occurs in the temporal lobe; it is this part of the brain that channels strong emotions into narratives. Flaherty also believes that changes in the temporal lobe are directly linked to both writer's block and the mysterious compulsion to express ourselves in writing. It is said that Fyodor Dostoyevsky had temporal lobe epilepsy—as did Lewis Carroll, who may also have suffered from hypergraphia. Either way: none of the remote diagnoses performed on Lewis Carroll have so far explained why his ailments took the form of an absurd and fantastic wonderland, or how this tormented and disease-ridden man was able to produce a creative universe that has captivated millions of people. Whatever it is, it can no more be reduced to a diagnosis than Van Gogh's sunflowers can be understood solely as a result of the artist getting heatstroke. Nor is death a conclusive explanation or recipe for creativity; it is just a condition for life that has a powerful effect on all who experience it. Death, perhaps more than anything else, creates new, empty spaces where something once was.

For Marte Spurkland, losing her father was like the floor of her house collapsing, leaving her floating midair. When struggling with such an enormous loss of meaning, all of our abilities to create meaning and coherence are put to the test. We look for connections and something to grab on to while we're falling, like Alice falling to Wonderland far below. Our culture has few, if any, gods—no fixed points. When faced by death, we have nothing to cling to but each other, to fragments of meaning, like Alice grabbing a jar of marmalade on her way down the rabbit hole only to find it empty. When faced by death, we can become highly creative and desperate for meaning.

Marte Spurkland developed magical thinking. Like the alchemists of the sixteenth century, Spurkland looked for secret connections in the universe and made her own strange deductions—that if she were successful, it would make her father's illness worse; if she thought negatively, it would make her father worse; if she forgot to wear a particular item of jewelry, that too would make her father worse.

"I knew that I'd slipped into a form of madness, an unstoppable crisis mode, that didn't really subside until now, now that the book is finished," she says.

This compulsion began with her father's diagnosis and still persisted when we met, one week after the book was launched. Grief had forced the book into existence.

There are so many books created by grief. Joan Didion wrote about her husband's death in *The Year of Magical Thinking*; Alison Bechdel made the graphic novel *Fun Home* about her father's suicide. *H Is for Hawk* describes how practical tasks (in addition to writing a book, of course) helped the author through an extremely difficult period. Ole Robert Sunde started two novels over the grief of his wife's death. The journalist Trude Lorentzen wrote a book about her mother's suicide, *My Mother: A Mystery*. Anna Fiske described a friend who takes his own life in her illustrated story *Danse på teppet* (Dancing on the rug). The death of Karl Ove Knausgård's father starts the first novel in his series My Struggle. In fact, there are so many books that start with someone dying mysteriously that it has become a genre of its own—crime fiction; you may have heard of it.

The Odyssey and *The Iliad* are two masterpieces created out of war and death; Orpheus searches for Eurydice in the Kingdom of the Dead; Dante went to Hell and met the dead; Goethe's Faust met the Devil; Mary Shelley's Frankenstein created a monster out of body parts. But death has inspired far more than books. Entire civilizations have been fixated on death and the afterlife, building pyramids and temples, churches and mosques, synagogues and memorials.

Lewis Carroll—the son of a priest, the great-grandson of a bishop, and a church deacon himself—must have known what he was writing about: Alice opening the door to a wonderland and finding a world beneath our own world where everyone is mad; she is descending into Hades, Hell, the Kingdom of the Dead. It is scary and funny—and, at the time, was the complete opposite of a children's book.

And that's actually how this book started. I experienced death. It is the reason I am writing: Vera stood face-to-face with death, and did so without losing any of her composure. When she was diagnosed with cancer, she didn't complain or become bitter; she just said that she had "a little bit of cancer," laughing with typical gallows humor. And as the end approached, she threw a party for all of her friends, where she lay on a sofa surrounded by flowers and balloons and the rhythm of wild Balkan music, and said goodbye to all the people she loved. At the end of the night—after over a hundred people had cried and laughed and drunk and danced in her honor—Vera was carried out by an ambulance crew. To the children who read her newspaper column, she wrote one final message:

> When you read this, I will be dead. Unfortunately, I have been living with cancer for over four years. The cancer I have is a type people in Norway don't normally die from, but I have been very unlucky. Anyway, I am grateful for the life I have lived, and I am not especially afraid of dying either.
>
> I often try to assure my grief-stricken friends that I'm honestly grateful to have experienced being seriously ill. Because it's not been *all* horrible. Quite the opposite. I have received so much warmth, so many hugs, such a huge amount of love, that it has felt like an enormous gift. Not everyone receives that kind of love in their life. The only thing that could top it, apart from getting well, of course, is that you all realize how great a gift life is, and that you turn to those close to you and make sure they understand how much they mean to you.

One of my dearest and most amazing friends died, and after she had gone, I knew I had to write a book about creativity, the thing at the very core of Vera's being. It is she who is Alice in this book. She went down into the black hole, and never came back. And I want you to know that she was here. Vera existed, someone like her existed: wonderful, creative, kind Vera was here, in this world, and then she was gone. Hilde raised this stone in memory of Vera.

After she died, my world turned black. It had become mad, a genuinely dark and scary wonderland. Because how could there be a world without Vera in it? It didn't make sense. The pattern had unraveled, and my lovingly stitched-together world had been reduced to a collection of loose, unsightly pieces. Out of habit, I would set off absent-mindedly for the hospital before remembering that Vera's old room was now occupied by a new cancer patient, one I didn't know. I can still feel an imprint of her in my life, her dry, slender hand in mine at the hospital, her cheek against mine when we hugged. I remember how she scolded me if I didn't undo the knitting mistakes, how she would laugh at a joke, how she always spurred me on, and made my life better. My memories of her are so real and vivid, and yet so fleeting and impossible to hold on to. The most tangible thing I can remember her by is a sweater she knitted for me at the hospital. When I wear it, I remember her hands moving rhythmically in the white hospital room and how the last thing we said to each other was "I love you."

I lost many things the year Vera died: I quit my job; we sold our apartment and bought a much smaller, more affordable one; and I was still suffering from the aftereffects of the concussion. We refurbished the apartment and moved in. It was chaos, and by Christmas, I was struggling to find a direction for my life.

It was then I started writing this book.

All I could do, it seemed, was to try to write a book that reminded me of everything that matters in life; it felt imperative. And what did I have to lose? I had nothing to lose anymore! I had to write about the Queen of Hearts and Cheshire Cats and weird inventions and the theory of relativity and Frankenstein's monster, about my child and my friend, about knitting and grief. I had to write about death, the sea, and love. I couldn't stop myself.

Even now, years later, I'll still get overwhelmed by the grief; it will come suddenly and I'll burst into tears. My daughter will always diligently attempt to comfort me.

"Don't be sad, Mama," she told me when things were at their darkest. "Everyone has to die. And then new people come."

She's absolutely right, of course—new people come, and new meaning comes. When someone dies, the world shifts a little, and everything changes. Something is born—an energy, a desperate need to understand what has happened, or at least to remember it: Mnemosyne, the goddess of memory, was the mother of nine muses for a reason. We raise memorial stones and build bridges. Creativity is not just a product of excess and abundance; it is also a pure necessity. To live is essentially to be creative. We search for meaning and cohesion, and ideas and notions emerge when we least expect them; we create patterns and see connections as a way of coping with the darkness we encounter. *Believing is weaving*. Vera would have liked my new slogan.

This book was born of grief—and is now being released into the world. It is a time capsule, sent into the future full of hope—and like many books, a bridge and a gift to someone who didn't ask for it. But then it strikes me: What if people don't read books in the future? What if we raise all these memorial stones, and years from now there's no one left to be reminded of anything?

8½ | The Dodo's Lament

OR: A WALK IN THE FUTURE FOREST.

"Wake up, Alice dear!" said her sister.
"Why, what a long sleep you've had!"

HUMANKIND HAS SUFFERED catastrophes before throughout history, but the catastrophe we are facing now is so much bigger and harder to comprehend. We have always been interested in creating doomsday scenarios, like the Great Flood or the Revelation of St. John. We have always wondered how the world might be when everything we know is gone. But this is no longer just about apocalypses. Over the last few years, most of the realistic stories about the future have concerned our planet being permanently changed. We live in the "Anthropocene," facing the terrifying prospect of a climate crisis. The novelist Maja Lunde's way of coping with this has been to combine her climate anxiety with the joy of writing.

"When I was a child, we had an anti-nuclear-weapons poster hanging over the kitchen table, and I've always been passionate about nature. So my writing is very much based on my own concerns."

Lunde finds both comfort and hope through story writing: it allows her to play out the very worst scenarios, but she can also explore the brightest rays of hope.

"I do this because I love to write, and I need to write. People ask, 'Doesn't it make you sad and frightened when you write about the climate and nature crisis?' And the answer is yes, of course; my anxiety grows the more I experience the stories. And although I sometimes want to escape from my own words—not writing *at all* makes me even sadder. My stories allow me to grieve, and they bring me comfort. Writing gives me the courage to explore both my darkest thoughts and hope itself. I write because I can, and because I need it," she says.

As I write this chapter, the latest UN report has determined that one million species are in danger of extinction. And when I read the list, it strikes me that many of the strange and beautiful and funny animals that appear in *Alice's Adventures in Wonderland* could be wiped out in the next few years. The only animals from Alice's world that are common right now, and will probably continue to be so, are white rabbits and cats.

But one animal in the book was already extinct when it was written; the peculiar-looking dodo, who so kindly starts a competition where everyone becomes the winner. The bird supposedly represents Lewis Carroll himself, whose troublesome stuttering often made him introduce himself as "Do-do-Dodgson" (his real last name). By the time Carroll first depicted this comical bird, it was already mythical, wiped out from the island of Mauritius by European explorers at the end of the seventeenth century. But this was the tentative start of human-driven species extinction. Now the other animals Carroll wrote about could find themselves following in the dodo's footsteps—flamingos, walruses, turtles, oysters, and hedgehogs are all on the IUCN Red List of Threatened Species. What if my daughter reads *Alice's Adventures in Wonderland* in a few years' time and thinks that *everything* in it is fiction?

When the climate crisis hits, we may need creativity more than ever—to rescue species, to survive extreme weather, to house climate refugees, and to produce food in completely different environmental conditions. And while the internet is controlled by rich moguls, and algorithms are threatening to steal our jobs, we will need creativity. But being creative means taking detours, digressions, and daydreams—and listening to our intuition—the exact opposite of what we actually want to do in a crisis: to run around, waving our arms and screaming, which is definitely not the best thing to do, no matter how tempted we are to resign ourselves to a sense of it being too late.

"We haven't talked enough about how the future may unfold, and we haven't taken into account the psychological aspects of envisioning these future scenarios," says Stanford professor Chris Field, a climate scientist who once sat on the Intergovernmental Panel on Climate Change (IPCC).

Doomsday visions are the most simple to portray, but they also have the most crippling effect on the vast majority of people: If there's no hope for the planet anyway, why should we do anything about it? Many of these gloomy outlooks predict more international conflicts arising from the struggle for resources, increasing numbers of refugees, a million species becoming extinct, and rising sea levels. Yet Field insists there is also reason for optimism.

"The ongoing changes in the climate will inspire development. New solutions for green energy will be pushed forward, bringing positive changes, especially for the world's poorest. Climate change may open up possibilities that will make the world a better place, creating stronger and more vibrant communities. This could be our chance to make some substantial changes," says Field.

In his essay collection *Romutvidelser* (Room expansions), the author Thure Erik Lund points at the huge paradoxes associated with a climate emergency—caused by the industrialization and consumerism that have made our lives so comfortable, and a middle class that valued art, literature, and freedom from authority and traditions. All good things at the time—but also one of the reasons for the problem we have now.

"European art is the mother of modernity. So it is therefore both the basis of the crisis and its savior. Because what do art and literature actually teach us now? To not worry about the unknown. That nothing is sacred. And to allow the consequences of nature to escalate. And keep moving. It is here that the climate crisis, our escalating technology, and literature come together, in one great communicative structure," Lund writes.

Literature's ability to explore the unknown is crucial, he believes. Like Chris Field, he thinks we have an opportunity to create something entirely different. The nightmare scenario is to carry on as before, allowing all the dysfunctional structures that brought us this problem to continue unabated.

In a way, we are still living in the age of the alchemists, who in the sixteenth century dreamed of having power over the entire world, and of manipulating nature for the benefit of humankind. All the world's forces should be subject to the godlike alchemist. Without knowing it, these delusional magicians paved the way for modern science, not with their pseudoscience and pointless experiments with lead and mercury, but because of the way they viewed the abilities of mankind. The alchemists saw a powerful, creative ruler of the world, an artist and scientist, impervious to religious authority. Now scientific thinkers and engineers, who control nature using figures and calculations, have taken the alchemists' place, and have contributed to incredible amounts of growth throughout the world. We perhaps need to forfeit some of this growth.

There are many people who now think it is time to approach the future differently. No one knows what the outcome will be, and our most dystopian thoughts and ideas about the future could actually create a dystopia. So if the stories we share about the future determine how it is formed, we need to be very careful which scenarios we choose.

"There are at least two schools of thought within climate communication. One highlights the need to inspire action and offer hope. The other believes that we should have sounded the alarm earlier and louder, and need to beat a very loud drum to really wake people up. Greta Thunberg, who has done an incredible job engaging young people, fits this description when she talks about wanting a climate panic," says Bård Vegar Solhjell, a former top politician, environment minister, and recently retired secretary-general of the World Wildlife Fund (WWF) in Norway.

Solhjell does not think panicking is a good idea. When we scream the word "crisis," we collectively go to into *sympathicus*, and we know what happens then: we get a racing pulse, clammy hands, a tummy ache, and a weak immune system. Some people will want to give up and "play dead"; others will try passive or active repression. Some will go into attack mode. What is certain is that none of this leads to good, long-term, and creative solutions.

He thinks we need to talk about opportunities, not catastrophes.

"We have to look at the psychology of this. What makes us act? We need to talk about the positive aspects. We need to talk about creativity, because this will be important in terms of how we deal with the climate crisis. Waiting for new technology to emerge and solve these problems is completely unnecessary, because virtually all the technology we need already exists—technology like 'stop cutting down trees,' or just 'start planting trees.' What we need more than new technology is human courage and creativity, and an ability to think differently," he says.

We are sitting in a vegan restaurant. Solhjell is not a vegetarian, but he eats here occasionally—and he stresses that our ability to make great vegetarian food is one creative solution that will have a beneficial effect on the future of the environment. For example, producing a soy burger requires "only" 42 gallons of water, compared with 621 gallons required for a hamburger. Cattle farming also accounts for huge emissions of methane, a gas that is even heavier than CO_2 and that exacerbates the greenhouse effect. So it's quite obvious which of these is better for the environment. Animal-based food accounts for 14.5 percent of all emissions, and slightly more than 20 percent of the emissions come from agriculture and changes in land use. We do not have to stop eating meat completely, but in the summer of 2018, the Intergovernmental Panel on Climate Change called on people to reduce their meat consumption and replace it with plant-based food. That is certainly no problem if it's really tasty, like the food we're eating now. Or like, perhaps, the most revolutionary thing to happen in the world of vegan food—something made by genuinely creative people who have succeeded in conjuring up the food of the future.

When Patrick O. Brown, a biochemistry professor at Stanford, took a sabbatical year (note that it was during a sabbatical!), he had an idea for reducing global meat consumption—because of the associated high methane emissions. And he found that the best way to do that was to make the perfect plant-based burger, a burger that would make people stop eating meat voluntarily. Using his knowledge of biochemistry, he was able to analyze the flavors a burger should contain and replicate them at a molecular level. The result was the Impossible Burger, which was launched in 2016, complete with a light pink, raw-looking texture inside.

Across the table, Bård Vegar Solhjell tucks into his Buddha bowl, although it's the Impossible Burger he ate recently that really made an impression on him. Solhjell believes it's without a doubt one of the best burgers he has ever eaten. Companies like Impossible Foods and Beyond Meat have transformed the vegetarian food market, producing plant-based burgers that are indistinguishable from real meat. And there are many other creative projects offering solutions to climate issues. For several years, the Oslo-based environmental group Bellona has been working on the Sahara Forest Project, which has transformed parts of the Sahara Desert into cultivatable land. The Canadian-Jordanian architect Abeer Seikaly has designed and made a tent that collects water and solar energy, which, due to the unfortunate increase in climate refugees, will be very much needed. There is also the EU quota system, which is an economically creative way of solving climate problems. And the tech billionaire Bill Gates is currently involved in a project developing a type of "wall" that extracts CO_2 from the air and converts it into energy—a kind of synthetic forest.

And recently David Vaughan at the Mote Marine Laboratory in Florida discovered by chance that he could make coral grow forty times faster in the laboratory than in the wild, and he is now aiming to plant millions of corals around the world over the next year. It is a project that offers hope to the world's coral reefs, which are so essential to the oceanic climate.

"But with an increase of two degrees, all the coral reefs will die because they can't survive at that temperature," says Solhjell. "Many large and irreversible climate changes will occur if we merely focus on stopping the warming at 2 degrees, instead of actually meeting the 1.5-degree target. I'm not going to lie to you—what's happening now is really serious, it's hard to process, and it makes you panic. But I'm an experienced optimist, and I become optimistic when I think about history: humankind has proven itself extremely capable of dealing with problems of our own making, and we've succeeded in changing course before. We've been in trouble many times in the past, in the last hundred years alone, and come out of it unscathed."

A reminder of some of what's happened since 1900: humankind has suffered two world wars that caused mass devastation and killed about 100 million people; it has also endured several major pandemics, such as the Spanish flu of 1918–1920, which also killed about 100 million people in the space of only two years. And *yet* we have sent people to the moon; we have invented the computer and the internet, cars and planes, telephones and television; we have developed completely new and revolutionary theories about the universe, mapped the human genome, discovered penicillin, implemented vaccine programs, carried out successful heart transplants, and started to understand the human brain. We are also, as I write, battling a new global pandemic, which has claimed millions of lives and led to the biggest vaccination effort the world has ever seen. Despite the incredible challenges we have faced, we are still here, in greater numbers, and better technically equipped, than ever before.

"Where solutions are concerned, there have been huge developments; the creativity is in full swing. And in a rational sense, we have every chance of limiting ourselves to the 1.5-degree target, or staying as far below 2 degrees as possible. The human and economic consequences of exceeding that would be so huge—why wouldn't we do it?" says Solhjell.

Humankind, in all its creative diversity, is sitting on both the problem and the solution. Mostly because, in practice, it's so difficult to define what we mean by "we." If we are going to solve a problem that affects us all, we all need to be united about it. But as far as climate issues are concerned, our strange and creative brains tend to set agendas or make plans that are conflicting, detrimental, or simply irrelevant. We are doing our own thing, all over the world. How can all humankind save all humanity, together, when we are all so incredibly different and unmanageable?

"I don't believe in climate dictatorships, but I've noticed that many people now are beginning to talk that way. It's of course tempting to think that the end justifies the means. But it would be wrong, because there's no one definitive way forward; we need dynamism and development and a collision of ideas. Besides, dictatorships aren't known for being particularly climate-friendly," says Solhjell before hurrying on his way, hopefully to save the planet—because from what I understand, it's getting a bit late for a U-turn.

"I'm used to working to time limits; it's how I work best. When I write, my productivity increases the closer I get to the deadline. I'll never be satisfied until I've worked hard for something," he says.

Maja Lunde is also cautiously optimistic that we will succeed in changing the world. It's not something that necessarily motivates her writing, but it is certainly the result: the stories in her climate quartet grab the reader emotionally and show them what the world might become if we fail to limit global warming to 1.5 degrees. The ideas for her books about species extinction and melting glaciers emerge from what she reads, and thinks about, daily. She finds the work deeply meaningful.

"My novels have given me an opportunity to talk about the most important theme of our time, and a voice in the climate debate. And it's a much-valued opportunity, because the fact that it feels meaningful is like a cure for the anxiety and hopelessness. In the periods between writing, I'm also able to reflect on what role literature, and maybe my own books, can play in light of this crisis."

Lunde believes that literature can change us, that it can help us understand climate issues in a more profound and emotional way—by making the scenarios more vivid.

"We need to feel anxious and panic-stricken in order for us to want to change. If we are to understand the significance of the crisis, we have to acknowledge it. Fiction can play a part in this acknowledgment, both for those writing it and for those reading it. Literature can provide comfort and spur people into action. I often get feedback from readers who have changed their lives in both large and small ways after reading my novels. And it makes me feel grateful. It's perhaps no coincidence that in Germany, where *The History of Bees* has sold more copies than anywhere else, they have initiated a large-scale project to save their bees," says Lunde.

Although Lunde doesn't base her writing on any overly preachy message, her climate quartet has changed the world anyway. Her books have been translated into thirty languages and have touched millions all over the world. Far more people have read her novels than have studied the UN climate reports.

I feel like our culture is a kind of network, like a spiderweb full of tiny raindrops, suspended between all sorts of little trees, and that it's possible to make this fragile network move. One person can create large vibrations. One little Greta can make the whole world listen, especially since we're all connected by the seeing telephones first envisioned by John Logie Baird and can communicate with anyone, anywhere, all of the time.

We are connected through our bodies and our brains; we are connected through history, through language, through the nature around us, and we can change each other in ways we are not even aware of.

Anne Beate Hovind is working on a project with the potential to change people who are not yet born. It is a beautiful June day and, with the city sparkling in the sun below, we're off on a little hiking trip. I've recently started to enjoy hiking a lot more.

"A lot of people might say what I'm talking about is banal, but there's a reason why the world's media listens to me when I talk about cathedral thinking and future generations," she tells me on our way into the forest.

"I'll often work with the implementation of a project and be very focused. But when I'm exploring something new and don't have anything to go on, I work very openly and, as a rule, don't allow myself to be directed by fear. If I did, then I wouldn't learn anything. We have to help each other to not be fear-driven," she says.

If art gives us space to think, to step out of ourselves and gain new perspectives, it will also be important to cultivate art in the years to come. (Speaking of not waving your arms and running around screaming—the opposite of that is to experience art.) This could be how we turn the situation, and the climate crisis, into a bright future for the coming generations. In 2011, Hovind invited the artist Katie Paterson to Norway with a view to working on a new project. Paterson spent a week alone in a cabin up in the mountains, then called Hovind and said that she finally knew what to do: she had an idea, something she had refined from an earlier draft. It would be a project with such an epic time span that she would be long gone before it came to fruition.

The Future Library's first author, Margaret Atwood, submitted her manuscript in 2014. Since then, several other world-class authors have submitted theirs, at an annual handover ceremony that takes place in the forest. All this from an idea that began on a train to Whitstable.

"I am interested in time. A lot of my work is very melancholy, some of it is absurd. But deep down it addresses the world of the ephemeral, how relative we are to the rest of the universe," Paterson says.

One of the themes of Paterson's work is the climate crisis, and the smallness of humankind in the face of nature and the universe. Her art has involved distant stars and solar eclipses, which have made time an important factor: human lives are so short compared with the immense age of the universe.

"I find thinking about planets slightly comforting: we humans wouldn't be here were it not for an exploded star. We're all part of something much bigger; we're breathing now because of a colossal sequence of events millions of years ago. When you think about it, life is about the awe and wonder at how immensely small our lives are. That's the paradox: to be a human is such a tiny and such an incredible thing at the same time," she says.

Paterson's latest work, the glass pillars made of melted sand, explores the theme of desertification caused by climate change. As with Maja Lunde's book *The End of the Ocean*, water and water scarcity is an important theme for Paterson.

"Climate change is part of my work. One of the reasons all the books in the Future Library are stored on the top floor of Oslo's new public library is the idea that the Oslofjord may rise, so I wanted the books to be protected at the very top of the building. And the same goes for the trees that are supposed to become books in 2114; we don't know what might happen to them, whether they'll survive or be destroyed by insects or forest fires. The first obstacle and global issue we encountered while working with the Future Library was the COVID-19 pandemic, which prevented us holding the annual handover of manuscripts. So the problems the Future Library has faced show us how vulnerable we are," she says.

In the forest north of Oslo, Anne Beate Hovind—the Future Library's project manager—and I are looking for the saplings that will one day become Future Library books. Finally, after walking around for thirty minutes, we spot a sign pointing them out. While our tour companion, Zolo the dog, basks in the sun beside us, we sit on a couple of tree stumps to admire this new patch of forest. Although the contents of these future books will remain a secret to readers until well into the next century, the Future Library has already gained international fame, with people from all over the world coming to see the trees.

"It was a difficult project for me. In my linear and goal-oriented world, how was I supposed to convince anyone that we should create a library for books that cannot be read for one hundred years? It challenges everything we take for granted about place and time," says Hovind.

Now we're in the middle of the future forest. There are no books, just the possibility of them being here, yet there is an almost hallowed atmosphere. Above our heads, a wispy, heart-shaped cloud floats in the summer sky.

"With this project, we've started talking about how the present isn't just *now*. The present spans two hundred years—two generations back and two generations ahead. If you start thinking like that, you start acting differently. Your pulse rate goes down, time feels different. Time becomes a landscape," she says.

The authors—who at this point are Margaret Atwood, David Mitchell, Sjón, Elif Shafak, Han Kang, Karl Ove Knausgård, Ocean Vuong, and Tsitsi Dangarembga—know that they will never see the reactions or criticism of the novel they submitted. It will be different for the authors submitting novels closer to 2114.

"So far, the most important thing for the authors has been the annual ritual of submitting the manuscript and the process of writing for the future. What we are doing here is a profound act of trust. We are trusting that future generations will take care of the project and pass it on, and they, in turn, have to trust that what we've passed on to them is good," says Hovind.

We sit alone in the warm sun, surrounded by the smell of heather and blueberries, and a forest that is crawling with busy little insects. Our eyes follow a couple of ants carrying something many times their own weight, just as this project is being carried forward, full of hope for the future, environmental perspectives, and cathedral thinking.

"Today's youth are afraid, and they look to us, us adults, to welcome their engagement with today's issues. My generation probably felt like everything was getting better and better. But we can't say that about the next generation. The Future Library is a story about having hope for the future. We've lived through a brief moment in history when nature wasn't managed properly, when we didn't listen to the experiences of our elders. And this period will end, and we will become conservationists. It's a project that speaks to everyone, in all cultures; it is global and directly addresses our basic need for belonging and hope," says Hovind, while stroking Zolo's curly black coat.

Zolo pants and whines a little, and we realize that it's time to go, since dogs unfortunately have a low threshold for boredom. We set off on the path, with Zolo trotting happily beside us. The heart-shaped cloud has evaporated in the sky. Maybe not everything will be forgotten one hundred years from now, as Knut Hamsun said; in one hundred years, these books will be read in a world different from ours, where there are people who hopefully *do* still read books and have done the best they can to save the earth from the climate crisis.

On my trip to the forest, I think of Vera. I would love to have told her about all I've discovered since she died! I've found out that creativity is everywhere all the time; it is unstoppable and unbreakable, whether it's been facilitated or not. People get ideas while they are sleeping, while sitting on the toilet, on the train, on the bus, when scared, when happy, while tying their shoelaces, while walking, while surfing the internet (yes, even then!), and while they are thinking about something completely different. I would have told her about parasympathetic mode, where we can daydream as much as we want and smell raspberries and stroke cats, and play and dance and laugh, make friends and share stories around the fire. I wanted to tell her how important daydreaming is (she would have known that already) and how little respect we have for it, perhaps because it reminds us of something childish (something we pretend that we're not), and, as a result, we fill our days with important assignments and stress. We are efficient and busy and can't even sit on the bus without checking our work email. How did we become so busy? I would have talked to her about how commercial forces exploit our creativity, sponge off it, make it small and banal, or try to mold it into a particular shape. There are so many people wanting to entertain us that it's easy to forget that proper learning, real engagement, and the ability to go into your own thoughts are some of the most important ingredients for creativity, and for living a good life. I would have told her that I've started knitting again—that I'm knitting a sweater for my daughter—and that knitting is genuinely good for creativity, even though I still kind of hate it. She would have liked that.

Finally, I wanted to tell her about my rabbit hole. Yes, that's right! A rabbit hole in my brain! I needed to find out if my cycling accident had given me permanent brain damage, or if I was just going crazy, so I consulted no fewer than two experts.

"Many people find that their life collapses after a concussion or a minor head injury. It's not unusual to lose your job or end up divorced," said one of the neurologists I consulted.

A lot of this research is still unclear. Brain researchers are in the process of taking a more thorough look at concussion and what actually happens when we suffer a head injury, and I've participated in one of these research projects myself, at Oslo University Hospital. What they do know is that many people find they have become a little lost, that even a minor head injury can cause major changes in the lives of those who experience it. And here I was, having changed just about everything in my life—my home, my job—and I had also buried a beloved friend.

Both of the two experts I consulted declared me fit and healthy after carefully examining me. In their opinion, there was nothing wrong with me anymore, except for some mild tinnitus and the fact that I needed a pair of stylish prescription glasses.

"Look," said one neurologist, an enthusiastic guy in his mid-forties, pointing at a screen that allowed us both to study my MRI image. I could see the outline of my eyeballs and a bluish-looking cerebral cortex that curled along the edge of the picture.

"You have a hole here, in the frontal lobe. A tiny black dot on the right side of your brain. Do you see that? It's not harmful, nothing to worry about, but it's the only strange thing I can see. And no, it's not an early sign of Alzheimer's," he said reassuringly.

Alzheimer's, yes, it had crossed my mind: How did I ever manage to crash into a stone bridge on a straight path? Was there something wrong with me before it happened? But no, I now realized the only thing wrong with me was that I had checked my cell phone while cycling; I can still remember the annoying ping from my pocket just before the impact. I had just been stressed. It was that that almost killed me. And now it turned out that I had a rabbit hole hidden in my brain. That makes sense, I thought, as I staggered out of the neurologist's office. Of course it does.

Three years have passed since the crash that suddenly changed my life. Three years of crying and fear and sorrow and debt collection letters and house-moving chaos and sickness and death; three years where everything I took for granted evaporated like mist in a damp forest. I've always thought that I would be able to return to the way I was before, but I now realize that can never happen, not after everything I've experienced. Nothing ever stays the same. It is the only thing we are truly guaranteed in life. The world will never return to how it was.

And as my shoes crunch upon the gravelly forest trail, I suddenly remember *The Tempest*, one of Shakespeare's last pieces, written in 1611 when the world was unstable and like a never-ending carnival on the threshold of all the modernity we know today. Shakespeare described a sorcerer, ruling over a magical island with his innocent teenage daughter Miranda, whose life is turned upside down by a terrible storm. And there begins the play. I think about the approaching storm. I think about the sea and the music and Ariel, the spirit of the air who fills the stage with inspiration—I breathe in, I breathe out, as I walk in the woods and feel the warm breeze on my body.

"We are such stuff as dreams are made on, and our little life is rounded with a sleep," Shakespeare wrote in his play, and a few years later, he was dead.

But the play ends with Prospero sending his lovestruck teenage daughter Miranda out into the world, along with young

Ferdinand. They are about to correct all the mistakes their fathers made during their bitter conflicts and power struggles. Shakespeare arms them with four qualities: love, hope, resolve, and creativity.

Thank You

MANY THANKS TO all my sources who have given their time to this book, and all the friends who have spurred me on while it was being written. Special thanks to Maja Lunde, Anna Fiske, Siw Aduvill, and Peder Kjøs, who have been not only benevolent sources, but also the best friends one could have; they have looked after me, supported me, and been good conversation and sparring partners along the way. Thanks to the authors Caterina Cattaneo and Vilde Kamfjord, two of my heroes and favorite people, who have championed me, hugged me, and read my words. Thanks to Linda Lund Nilsson, who spurred me on from afar. Thanks to Eivor Vindenes, Tone Holmen, and Hedda Klemetzen in my reading club, "Mon Amour." Thanks to my good friend Marit Ausland, who I meet every weekend in the botanical gardens, and Bente Roalsvig, who made me go to yoga every Sunday. Thanks to Silje Hegg, Truls Petersen, and Ellisiv Horrell, who have been there for me, and to everyone else who has tried keeping me parasympathetic throughout the last years.

Thanks to Mina Adampour, who has made me think differently about trees. Thanks to Ivo de Figueiredo, who always makes me a little wiser, and certainly helped regarding this book. Thanks to Siri Hustvedt, who is my guiding star and role model, and who has shared her knowledge and time so generously. Thanks to Professor Gaute Einevoll, who gave his precious time to reading this book, and whose interdisciplinary engagement is so inspiring.

Thanks to Erik Møller Solheim, my editor, friend, and savior. Thanks to my family, not least to Matt, my husband and almost too meticulous translator, for all of your hard work transforming this book from Norwegian to English.

Thank you to all the wonderful people at Greystone for believing in me, and especially to James Penco, my relentless English editor. I also want to thank Ingvild Haugland Blatt and her whole team at Cappelen Damm Agency for selling this book abroad, and thereby making it available in English and a lot of other languages.

None of the people I'm thanking here are responsible for my mistakes, but if I've done something worthwhile, I hope they get credit for it.

Thank you, Vera, for everything you gave me, for being my friend, and for showing me all that creativity can be. You will always be in my heart.

Sources

Quotations not cited here come from interviews that the author personally conducted between October 2018 and March 2022. All online sources checked as of 2021.

Page viii. The opening quote of the book is from: Lewis Carroll, *Alice's Adventures in Wonderland* (London: Macmillan, 1865).

INTRODUCTION: I HIT THE WALL BY THE RIVER AKERSELVA

Page 3. About head injuries (in Norwegian): Anne-Kristin Solbakk, Anne-Kristine Schanke, and Jan Magne Krogstad, "Hodeskader hos voksne: diagnostikk ogg rehabilitering" [Head injuries in adults: Diagnostics and rehabilitation], *Psykologi*, September 1, 2008, https://psykologtidsskriftet.no/fagartikkel/2008/09/hodeskader-hos-voksne-diagnostikk-og-rehabilitering.

Page 3. American website about traumatic head injuries (TBI): "Traumatic Brain Injury: What to Know About Symptoms, Diagnosis, and Treatment," U.S. Food & Drug Administration, August 23, 2021, https://www.fda.gov/consumers/consumer-updates/traumatic-brain-injury-what-know-about-symptoms-diagnosis-and-treatment.

Page 4. About dopamine production and blows to the head: Y. H. Chen, E. Y. Huang, T. T. Kuo, J. Miller, Y. H. Chiang, and B. J. Hoffer, "Impact of Traumatic Brain Injury on Dopaminergic Transmission," *Cell Transplantation* 26, no. 7 (July 2017): 1156–68, https://doi.org/10.1177/0963689717714105.

Page 4. About hypergraphia: Alice W. Flaherty, *The Midnight Disease: The Drive to Write, Writer's Block, and the Creative Brain* (New York: Houghton Mifflin, 2004), 11, 13.

Page 4. About Maurice Ravel's head injury: A. Otte, P. De Bondt, C. Van de Wiele, K. Audenaert, and R. Dierckx, "The Exceptional Brain of Maurice Ravel," *Medical Science Monitor* 9, no. 6 (2003): RA134–39.

Page 5. About executive function and working memory: Hilde Østby and Ylva Østby, *Adventures in Memory: The Science and Secrets of Remembering and Forgetting* (Vancouver: Greystone Books, 2018), 7–8, 195–99.

Page 6. About network destabilization and overactive DMN after a head injury: Randall S. Scheibel, "Functional Magnetic Resonance Imaging of Cognitive Control Following Traumatic Brain Injury," *Frontiers in Neurology* 4, no. 8 (2017): 352, https://doi.org/10.3389/fneur.2017.00352.

CHAPTER 1: THE CHESHIRE CAT APPEARS

Page 9. This is how much time is spent watching TV in the US: "The Nielsen Total Audience Report: August 2020," Nielsen, August 13, 2020, https://www.nielsen.com/us/en/insights/report/2020/the-nielsen-total-audience-report-august-2020/.

Felix Richter, "The Generation Gap in TV Consumption," Statista, November 20, 2020, https://www.statista.com/chart/15224/daily-tv-consumption-by-us-adults/.

Page 11. The story and quotes of John Logie Baird, from his autobiography: *Television and Me: The Memoirs of John Logie Baird* (Edinburgh: Mercat Press, 2004), 50ff.

Page 11. About the history of the BBC: "John Logie Baird," BBC, https://www.bbc.com/historyofthebbc/research/story-of-bbc-television/john-logie-baird.

"John Logie Baird (1888–1946)," BBC, https://www.bbc.co.uk/history/historic_figures/baird_logie.shtml.

Lincoln Allison, "The BBC's Worst Mistake," *The Critic*, December 4, 2020, https://thecritic.co.uk/the-bbcs-worst-mistake/.

Page 12. About Nipkow: "Paul Nipkow," Baird Television, last updated April 11, 2021, https://www.bairdtelevision.com/nipkow.html.

Page 14. A lot of good ideas weren't actually very good; here is a collection: Julie Halls, *Inventions That Didn't Change the World* (London: Thames & Hudson, 2014).

Page 14. About Reichelt and other inventors who failed: Ola Vikås, *Oppfinnere som døde av sin egen oppfinnelse* [Inventors who were killed by their own inventions] (Oslo: Ljå Forlag, 2014).

Page 14. See the Flying Tailor's last flight: "Franz Reichelt's Jump off the Eiffel Tower (1912)," British Pathé, uploaded on July 27, 2011, YouTube video, 1:35, http://www.youtube.com/watch?v=FBN3xfGrx_U.

Page 17. Archimedes's story, among others: William B. Irvine, *Aha! The Moments of Insight That Shape Our World* (New York: Oxford University Press, 2015), 4–6.

Page 17. But did Archimedes really run naked through the streets of Syracuse? The website the Naked Scientists investigates: Alan Hirshfeld, "Archimedes: The Original Naked Scientist," The Naked Scientists, October 10, 2010, http://www.thenakedscientists.com/articles/science-features/archimedes-original-naked-scientist.

Page 17. What inspiration and ideas meant to the Greeks: E. R. Dodds, *The Greeks and the Irrational* (Berkeley: University of California Press, 1951), 64ff.

Page 19. Aristotle on melancholia, quoted in: Alice W. Flaherty, *The Midnight Disease: The Drive to Write, Writer's Block, and the Creative Brain* (New York: Houghton Mifflin, 2004), 32.

Page 20. About Ficino and Saturn's melancholy influence: Frances A. Yates: *Giordano Bruno and the Hermetic Tradition* (Chicago: University of Chicago Press, 1964), 130–68.

Page 20. About *The Tempest* and alchemy: Hilde Østby, "Under melankoliens stjerne. Shakespeares The Temmpest tolket I lys av magi og betydningen av Saturn i renessansen" [Under the star of melancholy: Shakespeare's *The Tempest* interpreted in the light of magic and the meaning of Saturn in the Renaissance], master's thesis, University of Oslo, 2000.

Page 20. This interpretation from 2018 links *The Tempest* to Ficino's student Pico della Mirandola: Michaela Krämer, "Through the Microscope: Understanding *The Tempest* in the Context of the Introduction to Pico della Mirandola's *Oration on the Dignity of Man*," 2018, http://www.thetempest.de/.

Page 22. About fast and slow thinking: Daniel Kahneman, *Thinking, Fast and Slow* (New York: Farrar, Straus and Giroux, 2011).

Page 22. Flow in perception, and "aha" moments: Jochim Hansen, Alice Dechêne, and Michaela Wänke, "Discrepant Fluence Increases Subjective Truth," *Journal of Experimental Social Psychology* 44, no. 3 (2008): 687–91, https://doi.org/10.1016/j.jesp.2007.04.005.

Page 22. The dark side of "aha" moments and why we believe in nonsense: R. E. Laukkonen, B. T. Kaveladze, J. M. Tangen, and J. W. Schooler, "The

Dark Side of Eureka: Artificially Induced Aha Moments Make Facts Feel True," *Cognition* 196 (2020): 104122, https://doi.org/10.1016/j.cognition.2019.104122.

Page 22. This is exactly what happens in the brain when we have an "aha" moment: Martin Tik, Ronald Sladky, Caroline Di Bernardi Luft, David Willinger, André Hoffmann, Michael J. Banissy, Joydeep Bhattacharya, and Christian Windischberger, "Ultra-High-Field fMRI Insights on Insight: Neural Correlates of the Aha!-Moment," *Human Brain Mapping* 39, no. 8 (2018): 3241–52, https://doi.org/10.1002/hbm.24073.

Page 23. Powerful emotions are triggered by even quite trivial "aha" experiences, and make them extra memorable: Rick Nauert, "Where 'Aha' Moments Reside in the Brain," Live Science, April 1, 2011, https://www.livescience.com/13529-insights-brain-region-aha-moments.html.

Page 25. Rolf Reber and Sascha Topolinski on "aha" moments: Sascha Topolinski and Rolf Reber, "Gaining Insight Into the 'Aha' Experience," *Current Directions in Psychological Science* 19, no. 6 (2010): 402–5, https://www.researchgate.net/publication/272160303_Gaining_insight_into_the_Aha_ Experience.

Page 28. The importance of dopamine is exaggerated (that dopamine gives a feeling of well-being and explains all addiction) and brain chemistry is more complex than is automatically assumed, writes neuropsychologist Vaughan Bell, who works at University College London: Vaughan Bell, "The Unsexy Truth About Dopamine," *Guardian*, February 3, 2013, http://www.theguardian.com/science/2013/feb/03/dopamine-the-unsexy-truth.

Page 29. On Einstein's thought experiment with the mirror, an idea he came up with when he was sixteen years old: John D. Norton, "Chasing a Beam of Light: Einstein's Most Famous Thought Experiment," faculty website, University of Pittsburgh, February 15, 2005, https://www.pitt.edu/~jdnorton/Goodies/Chasing_the_light/.

Page 29. The stories of Albert Einstein and Robert Feynman are told in: William B. Irvine, *Aha! The Moments of Insight That Shape Our World* (New York: Oxford University Press, 2015), 2–3, 132–36.

Page 30. Van Gogh quote from: Letter 689, from Vincent van Gogh to Theo van Gogh, September 26, 1888, http://www.vangoghletters.org/vg/letters/let689/letter.html.

Page 35. About how a little knowledge increases our need to learn something new radically: Celeste Kidd and Benjamin Y. Hayden, "The

Psychology and Neuroscience of Curiosity," *Neuron* 88, no. 3 (2015): 449–60.

Page 35. About the status of curiosity in the history of the West: Hans Blumenberg, *The Legitimacy of the Modern Age*, trans. Robert M. Wallace (Cambridge, MA: MIT Press, 1983), 229ff.

Page 42. Alexander Fleming's story is told in: William B. Irvine, *Aha! The Moments of Insight That Shape Our World* (New York: Oxford University Press, 2015), 141–42.

Page 44. A whole team of researchers claimed in 2013 that clutter was good for creativity: Kathleen D. Vohs, Joseph P. Redden, and Ryan Rahinel, "Physical Order Produces Healthy Choices, Generosity, and Conventionality, Whereas Disorder Produces Creativity," *Psychological Science* 24, no. 9 (2013): 1860–67.

Also see: Lea Winerman, "A Messy Desk Encourages a Creative Mind, Study Finds," American Psychological Association, *Monitor on Psychology* 44, no. 9 (October 2013): 12, https://www.apa.org/monitor/2013/10/messy-desk/.

Page 44. But in 2019, a new team of researchers concluded that a cluttered desk is not necessarily good for creativity: Alberto Manzi, Yana Durmysheva, Shannon K. Pinegar, Andrew Rogers, and Justine Ramos, "Workplace Disorder Does Not Influence Creativity and Executive Functions," *Frontiers in Psychology* 9 (2019): 2662, https://doi.org/10.3389/fpsyg.2018.02662.

Page 45. About Henry Molaison and memory research: Hilde Østby and Ylva Østby, *Adventures in Memory: The Science and Secrets of Remembering and Forgetting* (Vancouver: Greystone Books, 2018), 4ff.

Page 47. How creativity research is gaining momentum: Michael Bycroft, "The Birth of a Movement: Joy Paul Guilford and Creativity Research in American Psychology, 1950–70," paper for 23rd International Congress of History of Science and Technology, Budapest, July 2009.

Page 48. Quote from the man who invented "brainstorming" and who started an academy for creative thinking: Alex F. Osborn, *Applied Imagination: Principles and Procedures of Creative Problem-Solving* (New York: Charles Scribner's Sons, 1953), 240–41.

Page 48. Brainstorming has turned out to be not as good for creativity as Osborn presented it: Teresa Torres, "Why Brainstorming Doesn't Work (and What to Do Instead)," *Inc.*, February 4, 2016, http://www.inc.com/teresa-torres/why-brainstorming-doesnt-work-and-what-to-do-instead.html.

Page 48. About different types of creativity testing: Michael Bloomfield, "Can Creativity Be Scientifically Measured?," Medium, April 11, 2018, https://michaelbloomfield.medium.com/can-creativity-be-scientifically-measured-659ee7db8d9d.

Page 49. Quote from Lewis Carroll in: Jenny Woolf, *The Mystery of Lewis Carroll: Discovering the Whimsical, Thoughtful, and Sometimes Lonely Man Who Created* Alice in Wonderland (New York: St. Martin's Press, 2010), 266.

Page 49. The first two thousand copies of *Alice's Adventures in Wonderland* were donated to hospitals and various homes, and one of these ended up at auction recently: Alison Flood, "'Legendary' First Edition of Alice in Wonderland Set for Auction at $2-3m," *Guardian*, May 30, 2016, https://www.theguardian.com/books/2016/may/30/alice-in-wonderland-first-edition-christies-auction/.

Page 50. An early TV broadcast: "'The Man With a Flower in His Mouth' First TV Play ~ 1930," SilverScreenSurfer, uploaded on July 14, 2014, YouTube video, 9:00, http://www.youtube.com/watch?v=d-84EDrl8YU.

CHAPTER 2: THE MAD HATTER'S TEA PARTY

Page 51. Christopher Marlowe, *Doctor Faustus: With the English Faust Book* (Indianapolis: Hackett Classics, 2005). Also available at: https://www.gutenberg.org/files/779/779-h/779-h.htm.

Page 53. About "pruning the brain": Irwin Feinberg, "Why Is Synaptic Pruning Important for the Developing Brain?," *Scientific American*, May 1, 2017, https://www.scientificamerican.com/article/why-is-synaptic-pruning-important-for-the-developing-brain/.

Page 56. An experiment conducted using music teachers and their students showed that music teachers could not guess which of their students would become professional musicians as adults. Referred to in: Are Brean and Geir Olve Skeie, *Musikk og hjernen: Om musikkens magiske kraft og fantastiske virkning på hjernen* [Music and the brain: About the magical power of music and its amazing effect on the brain] (Oslo: Cappelen Damm, 2019), 123; and M. J. Howe, J. W. Davidson, and J. A. Sloboda, "Innate Talents: Reality or Myth?," *Behavioral and Brain Sciences* 21, no. 3 (1998): 399–407, https://doi.org/10.1017/s0140525x9800123x.

Page 57. Suzanne Moore, on lipstick and her own writing: Suzanne Moore, "Find a Room of Your Own: Top 10 Tips for Women Who Want to Write," *Guardian*, October 5, 2019, https://www.theguardian.com/lifeandstyle/2019/oct/05/find-a-room-of-your-own-top-10-tips-for-women-who-want-to-write/.

Page 57. About Freud and Vienna: Carl E. Schorske, *Fin-De-Siècle Vienna: Politics and Culture* (New York: Random House, 1981), 181ff.

Page 58. How damage to the temporal lobe can occur: "Temporal Lobes," Centre for Neuro Skills, https://www.neuroskills.com/brain-injury/temporal-lobes/.

Page 59. Brain waves explained:

Research from NTNU (Norway): Anne Steenstrup-Duch, "Meditasjon skaper bølger" [Meditation creates waves], Forskning.no, March 6, 2010, https://forskning.no/forebyggende-helse-ntnu-stress/meditasjon-skaper-bolger/863467/.

Jennifer Larson, "What Are Alpha Brain Waves and Why Are They Important?," Healthline, October 9, 2019, https://www.healthline.com/health/alpha-brain-waves/.

Page 59. Beeman's research on "aha" moments: John Kounios and Mark Beeman, "The Cognitive Neuroscience of Insight," *Annual Review of Psychology* 65 (2014): 71–93, https://doi.org/10.1146/annurev-psych-010213-115154.

Page 59. About the research of Beeman and Joy Bhattacharya: Lauren Migliore, "The Aha! Moment: The Science Behind Creative Insight," *BrainWorld*, April 6, 2020, https://brainworldmagazine.com/aha-moment-science-behind-creative-insight/.

Joy Bhattacharya has found that we know in advance whether a problem requires an "aha" experience or if it can be solved analytically. Bhavin R. Sheth, Simone Sandkühler, and Joydeep Bhattacharya, "Posterior Beta and Anterior Gamma Oscillations Predict Cognitive Insight," *Journal of Cognitive Neuroscience* 21, no. 7 (2009): 1269–79, https://doi.org/10.1162/jocn.2009.21069.

Page 61. Researchers in London came a little closer to understanding the inner critic: Nicola Davis, "Suppressing the Reasoning Part of the Brain Stimulates Creativity, Scientists Find," *Guardian*, June 7, 2017, https://www.theguardian.com/science/2017/jun/07/thinking-caps-on-electrical-currents-boost-creative-problem-solving-study-finds/.

Page 61. The two quotes by Caroline Di Bernardi Luft are from the article in the *Guardian*, cited above.

Page 61. Here is Bernardi Luft's research on the temporal lobe: Caroline Di Bernardi Luft, Ioanna Zioga, Nicholas M. Thompson, Michael J. Banissy, and Joydeep Bhattacharya, "Right Temporal Alpha Oscillations as a Neural Mechanism for Inhibiting Obvious Associations," *Proceedings of the National Academy of Sciences of the United States of America* 115, no. 52 (2018): E12144–52, https://doi.org/10.1073/pnas.1811465115.

Page 63. About the flow of electrical current through the brain, and creativity, a larger overview: Claudio Lucchiari, Paola Maria Sala, and Maria Elide Vanutelli, "Promoting Creativity Through Transcranial Direct Current Stimulation (tDCS). A Critical Review," *Frontiers in Behavioral Neuroscience* 12 (2018): 167, https://doi.org/10.3389/fnbeh.2018.00167.

Page 65. F. G. Barker II, "Phineas Among the Phrenologists: The American Crowbar Case and Nineteenth-Century Theories of Cerebral Localization," *Journal of Neurosurgery* 82, no. 4 (1995): 672–82, https://doi.org/10.3171/jns.1995.82.4.0672.

Page 66. About the frontal lobe: Elkhonon Goldberg, *The New Executive Brain: Frontal Lobes in a Complex World* (New York: Oxford University Press, 2009), 44.

Page 67. About traumatic head injuries and disorders of the executive function and DMN: Youngxia Zhou, Michael P. Milham, Yvonne W. Lui, Laura Miles, Joseph Reaume, Daniel K. Sodickson, Robert I. Grossman, and Yulin Ge, "Default-Mode Network Disruption in Mild Traumatic Brain Injury," *Radiology* 265, no. 3 (2012): 882–92, https://doi.org/10.1148/radiol.12120748.

Page 67. About injuries in the temporal lobe: "Temporal Lobes," Centre for Neuro Skills, https://www.neuroskills.com/brain-injury/temporal-lobes/.

Page 67. About the temporal lobe's many functions: "Temporal Brain Lobe," Human Memory, May 20, 2022, https://human-memory.net/temporal-brain-lobe/.

Page 68. Alan Baddeley on executive function: Alan Baddeley, "Exploring the Central Executive," *Quarterly Journal of Experimental Psychology* 49, no. 1 (1996): 5–28, https://doi.org/10.1080%2F713755608.

Page 69. The last hippie is described in Oliver Sacks's *An Anthropologist on Mars: Seven Paradoxical Tales* (New York: Alfred A. Knopf, 1995), 42ff.

Page 70. About alcohol use and creativity: William B. Irvine, *Aha! The Moments of Insight That Shape Our World* (New York: Oxford University Press, 2015), 279.

Page 70. Attempts at inhibitions and alcohol: Gabriela Gan, Alvaro Guevara, Michael Marxen, Maike Neumann, Elisabeth Jünger, Andrea Kobiella, Eva Mennigen, Maximilian Pilhatsch, Daniel Schwarz,

Ulrich S. Zimmermann, and Michael N. Smolka, "Alcohol-Induced Impairment of Inhibitory Control Is Linked to Attenuated Brain Responses in Right Fronto-temporal Cortex," *Biological Psychiatry* 76, no. 9 (2014): 698–707, https://doi.org/10.1016/j.biopsych.2013.12.017.

Page 71. About living healthy and the artist myth: Haruki Murakami, *What I Talk About When I Talk About Running*, trans. Philip Gabriel (New York: Alfred A. Knopf, 2008), 96.

Page 72. One Tracey Emin quote is from this article: Chris Hastings, "Tracey Emin: I Never Had Children Because I'd Resent Leaving My Studio for Them," *Daily Mail*, April 20, 2013, https://www.dailymail.co.uk/femail/article-2312355/Tracey-Emin-I-children-Id-resent-leaving-studio-them.html/.

Page 73. And one from this one: Jonathan Jones, "How Tracey Emin Is Giving Munch the Mother He Never Had," *Guardian*, January 2, 2020, https://www.theguardian.com/artanddesign/2020/jan/02/tracey-emin-interview-munch-mother-statue-oslo/.

Page 73. More about the male genius and what it's like for women who create: Marta Breen, *Om muse rog menn* [On muses and men] (Oslo: Spartacus Forlag, 2019); and Siri Hustvedt, *A Woman Looking at Men Looking at Women: Essays on Art, Sex, and the Mind* (New York: Simon & Schuster, 2016).

Page 76. About bipolar disorder and writing: Alice W. Flaherty, *The Midnight Disease: The Drive to Write, Writer's Block, and the Creative Brain* (New York: Houghton Mifflin, 2004), 14, 117–18.

Page 77. About Van Gogh's many illnesses: Kalyan B. Bhattacharyya and Saurabh Rai, "The Neuropsychiatric Ailment of Vincent van Gogh," *Annals of Indian Academy of Neurology* 18, no. 1 (2015): 6–9, https://doi.org/10.4103/0972-2327.145286.

Page 77. Quote from Vincent van Gogh: Letter 569, Vincent van Gogh to Horace Mann Livens, September or October 1886, http://www.vangoghletters.org/vg/letters/let569/print.html.

Page 78. Quote from Cary Grant: Michael Pollan, *How to Change Your Mind: What the New Science of Psychedelics Teaches Us About Consciousness, Dying, Addiction, Depression, and Transcendence* (New York: Penguin Press, 2018), 156–57.

Page 78. About microdosing of LSD: Vince Polito and Richard J. Stevenson, "A Systematic Study of Microdosing Psychedelics," *PLOS ONE* 14, no. 2 (2019): e0211023, https://doi.org/10.1371/journal.pone.0211023.

Page 78. About Steve Jobs, the tech industry, and LSD: Ryan Grim, "Read the Never-Before-Published Letter From LSD-Inventor Albert Hofmann to Apple CEO Steve Jobs," *HuffPost*, August 8, 2009, https://www.huffpost.com/entry/read-the-never-before-pub_b_227887.

Page 80. How placebo drugs can increase creativity: Liron Rozenkrantz, Avraham E. Mayo, Tomer Ilan, Yuval Hart, Lior Noy, and Uri Alon, "Placebo Can Enhance Creativity," *PLOS ONE* 12, no. 9 (2017): e0182466, https://doi.org/10.1371/journal.pone.0182466.

Page 80. About marijuana and creativity: V. Krishna Kumar, "Cannabis and Creativity," *Psychology Masala* (blog), Psychology Today, April 20, 2012, http://www.psychologytoday.com/us/blog/psychology-masala/201204/cannabis-and-creativity/.

Page 80. Luna Reyna, "Does Smoking Weed Really Make You More Creative?," *Dope Magazine*, January 31, 2018, https://tv.dopemagazine.com/cannabis-creative/.

Page 84. The new trend in London is improv theater: Stevie Martin, "London's Newest Craze Is Improv Comedy, Apparently," *Vice*, January 1, 2016, https://www.vice.com/en/article/3dx9wb/londons-newest-craze-improv-comedy-the-free-association.

Page 88. Toril Moi quoted in: Kristine Isaksen and Hans Petter Blad, eds., *Min metode: Om sakprosaskriving* [My method: About nonfiction writing] (Oslo: Cappelen Damm, 2019), 17.

Page 90. About rest and parasympathetic mode: Siw Aduvill, *Hvile: Alt vi vinner ved å la være* [Rest: All that you gain by letting go] (Oslo: Tiden Norsk Forlag, 2019).

Page 93. Audun Myskja, *Pust: Nøkkelen til styrke, helse og glede* [Breathe: The key to strength, health, and joy] (Oslo: J. M. Stenersens Forlag, 2019), 58.

Page 94. Women are more receptive to external motivation and criticism: Alice W. Flaherty, *The Midnight Disease: The Drive to Write, Writer's Block, and the Creative Brain* (New York: Houghton Mifflin, 2004); and Valentina Zarya, "This Is the Big Difference Between How Men and Women Respond to Feedback," *Fortune*, September 1, 2016, https://fortune.com/2016/09/01/gender-feedback-perception/.

Page 95. Aristotle and friendship: Aristotle, *Nicomachean Ethics*, book 8, trans. W. D. Ross, available at http://classics.mit.edu/Aristotle/nicomachaen.8.viii.html.

CHAPTER 3: PLAYING CROQUET WITH THE QUEEN OF HEARTS

Page 97. Stuart Dodgson Collingwood, *The Life and Letters of Lewis Carroll* (New York: The Century Co., 1899), location 1129.

Page 98. About how Michelangelo viewed marble as no more than a prison for the sculptures he released: Nils Parker, "The Angel in the Marble," Medium, July 8, 2013, https://medium.com/@nilsaparker/the-angel-in-the-marble-f7aa43f333dc.

Page 98. An example of Manichaean scripture: "Hymn About the Captivity of Light," Gnostic Society Library, http://gnosis.org/library/hymncapt.htm.

Page 99. Quotes from the essay "Modern Fiction" by Virginia Woolf: Virginia Woolf, *The Essays of Virginia Woolf*, vol. 4, 1925–1928, ed. Andrew McNeillie (London: Hogarth Press, 1986), 154.

Page 103. Story about Solomon Shereshevsky at the ice cream parlor: A. R. Luria, *The Mind of a Mnemonist: A Little Book About a Vast Memory*, trans. Lynn Solotaroff (Cambridge, MA: Harvard University Press, 1987), 82.

Page 103. The case of Solomon Shereshevsky is described in: Hilde Østby and Ylva Østby, *Adventures in Memory: The Science and Secrets of Remembering and Forgetting* (Vancouver: Greystone Books, 2018), 12–14, 201–2.

Page 104. Arthur Rimbaud's poem: Arthur Rimbaud, "Vowels," trans. Paula Ayer, 2023.

Page 105. Research on synesthesia and creativity: Jamie Ward, Daisy Thompson-Lake, Roxanne Ely, and Flora Kaminski, "Synaesthesia, Creativity and Art: What Is the Link?," *British Journal of Psychology* 99, no. 1 (2008): 127–41.

Page 109. Research on bilingualism and our understanding of time: E. Bylund and P. Athanasopoulos, "The Whorfian Time Warp: Representing Duration Through the Language Hourglass," *Journal of Experimental Psychology: General* 146, no. 7 (2017), 911–16.

Page 110. About *The Hero With a Thousand Faces*, by Joseph Campbell, and *Star Wars*: Kristen Brennan, "Joseph Campbell," Star Wars: Origins, http://www.moongadget.com/origins/myth.html.

Page 111. Thomas Suddendorf in: Hilde Østby and Ylva Østby, *Adventures in Memory: The Science and Secrets of Remembering and Forgetting* (Vancouver: Greystone Books, 2018), 240.

Page 111. About the difference between the story and how it is told: Seymour Chatman, *Story and Discourse: Narrative Structure in Fiction and Film* (Ithaca, NY: Cornell University Press, 1978).

Page 112. Wisława Szymborska, "Could Have," trans. Clare Cavanagh and Stanisław Barańczak, *Map: Collected and Last Poems* (New York: Houghton Mifflin Harcourt, 2015), 155.

Page 113. About mirror neurons: Richard Cook, Geoffrey Bird, Caroline Catmur, Clare Press, and Cecilia Heyes, "Mirror Neurons: From Origin to Function," *Behavioral and Brain Sciences* 37, no. 2 (2014): 177–92, https://doi.org/10.1017/S0140525X13000903.

Page 113. More about mirror neurons: Fabrizio Mafessoni and Michael Lachmann, "The Complexity of Understanding Others as the Evolutionary Origin of Empathy and Emotional Contagion," *Scientific Reports* 9, no. 1 (2019): 5794, https://doi.org/10.1038/s41598-019-41835-5.

Page 113. About mirror neurons and literature: Gerhard Lauer, "Going Empirical. Why We Need Cognitive Literary Studies," *Journal of Literary Theory* 3, no. 1 (2009): 145–54, https://doi.org/10.1515/JLT.2009.007.

Page 115. Kazuo Ishiguro's Nobel speech is published as a book: Kazuo Ishiguro, *My Twentieth Century Evening and Other Small Breakthroughs: The Nobel Lecture* (New York: Alfred A. Knopf, 2017), 37–38.

Page 115. Hilde Østby, *Leksikon om lengsel* [Encyclopedia of love and longing] (Oslo: Tiden Forlag, 2013), 240.

Page 116. A large study of 631 test subjects who had depression showed that bibliotherapy didn't really have any great long-term effect on depression: Sina Müller, Paul Rohde, Jeff M. Gau, and Eric Stice, "Moderators of the Effects of Indicated Group and Bibliotherapy Cognitive Behavioral Depression Prevention Programs on Adolescents' Depressive Symptoms and Depressive Disorder Onset," *Behaviour Research and Therapy* 75 (2015): 1–10, https://doi.org/10.1016/j.brat.2015.10.002.

Page 116. About the implied author and the implied reader: Wayne C. Booth, *The Rhetoric of Fiction* (Chicago: University of Chicago Press, 1961).

Page 116. About books as friends: Wayne C. Booth, *The Company We Keep: An Ethics of Fiction* (Berkeley: University of California Press, 1988).

Page 120. Quote from Siw Aduvill's *Hvile: Alt vi vinner ved å la være* [Rest: All that you gain by letting go] (Oslo: Tiden Forlag, 2019), 170.

Pages 120, 122. Haruki Murakami on finding his own flock and retiring, and on how he writes with direction: Haruki Murakami, *What I Talk About When I Talk About Running*, trans. Philip Gabriel (New York: Alfred A. Knopf, 2008), 37, 77, 80.

Page 120. About how cultural experiences counter depression, study done by University College London: Daisy Fancourt and Urszula Tymoszuk, "Cultural Engagement and Incident Depression in Older Adults: Evidence From the English Longitudinal Study of Ageing," *British Journal of Psychiatry* 214, no. 4 (2019): 225–29, https://doi.org/10.1192/bjp.2018.267.

Page 122. About how professional musicians are in both DMN and the executive function when they improvise: Psyche Loui, "Rapid and Flexible Creativity in Musical Improvisation: Review and a Model," *Annals of the New York Academy of Sciences* 1423, no. 1 (2018): 138–45, https://doi.org/10.1111/nyas.13628.

Page 123. The art of procrastination is described in: John Perry, *The Art of Procrastination: A Guide to Effective Dawdling, Lollygagging and Postponing* (New York: Workman Publishing, 2012), 19.

Page 125. Quote about the perfect first sentence taken from: Albert Camus, *The Plague*, trans. Stuart Gilbert (New York: Vintage, 1991), 104–5.

CHAPTER 4: WONDERLAND

Page 128. Doris Lessing on sleeping: Doris Lessing, "Doris Lessing: A Room of One's Own," *New Statesman*, March 27, 2013, https://www.newstatesman.com/long-reads/2013/03/room-ones-own/.

Page 129. Haruki Murakami, *What I Talk About When I Talk About Running*, trans. Philip Gabriel (New York: Alfred A. Knopf, 2008), 4.

Page 130. Lars Svendsen, *A Philosophy of Boredom*, trans. John Irons (London: Reaktion Books, 2005).

Page 132. Quote from: Dr. Seuss, *Oh, the Places You'll Go!* (New York: Random House, 1990), 26–27.

Page 132. Søren Kierkegaard on boredom in the Gilleleje manuscript from the summer of 1835: Letter written August 1 in Copenhagen, AA12 (in Danish), http://sks.dk/aa/txt.xml?hash=ss26&zoom_highlight=bundl%C3%B8se+hav#ss26.

Page 133. Quote from: Samuel Beckett, *Waiting for Godot* (New York: Grove Press, 1954), 53. Quote is taken from act 2, and can also be read here: https://www.samuel-beckett.net/Waiting_for_Godot_Part2.html.

Page 134. Americans watch 197 minutes a day on average: Julia Stoll, "Time Spent Watching Television in the U.S. 2019–2023," Statista, February 14, 2022, https://www.statista.com/statistics/186833/average-television-use-per-person-in-the-us-since-2002/.

Brits watch 192 minutes a day on average: Julia Stoll, "Minutes per Day Spent Watching Broadcast TV in the UK 2010–2020, by Age," Statista, August 25, 2021, https://www.statista.com/statistics/269918/daily-tv-viewing-time-in-the-uk-by-age/.

Canadians watch 28.7 hours a week, or approximately 180 minutes a day on average: Julia Stoll, "Weekly Time Spent Watching TV in Canada 2020–2021, by Age Group," Statista, April 20, 2022, https://www.statista.com/statistics/234311/weekly-time-spent-watching-tv-in-canada-by-age-group/.

Page 135. People would rather be electrically shocked than be bored for fifteen minutes: Timothy D. Wilson, David A. Reinhard, Erin C. Westgate, Daniel T. Gilbert, Nicole Ellerbeck, Cheryl Hahn, Casey L. Brown, and Adi Shaked, "Just Think: The Challenges of the Disengaged Mind," *Science* 345, no. 6192 (2014): 75–77.

Page 135. Pessoa quoted in: Lars Svendsen, *A Philosophy of Boredom*, trans. John Irons (London: Reaktion Books, 2005), 19.

Page 136. Manoush Zomorodi, *Bored and Brilliant: How Spacing Out Can Unlock Your Most Productive and Creative Self* (New York: St. Martin's Press, 2017), 19.

Page 138. Quotes by Marcus Raichle taken from an article in *Dyade* (Norwegian only): "Minner om fremtiden og vindskjeve furuer" [Reminiscent of the future and windswept pines], *Dyade*, March 2014, https://dyade.no/tidsskrift/dyade_2014_03_tankestroemmens_tidsalder/minner_om_fremtiden_og_vindskjeve_furuer.

Page 138. Also see this article by Marcus Raichle about the discovery of the DMN: Marcus E. Raichle, Ann Mary MacLeod, Abraham Z. Snyder, William J. Powers, Debra A. Gusnard, and Gordon L. Shulman, "A Default Mode of Brain Function," *Proceedings of the National Academy of Sciences of the United States of America* 98, no. 2 (2001): 676–82.

Page 138. We (Hilde Østby and Ylva Østby) also wrote about the DMN in: Hilde Østby and Ylva Østby, *Adventures in Memory: The Science and Secrets of Remembering and Forgetting* (Vancouver: Greystone Books, 2018), 247–48.

Page 139. Marcus Raichle and the discovery of the DMN: Marcus E. Raichle, "The Brain's Default Mode Network," *Annual Review of Neuroscience* 38 (2015): 433–47, https://doi.org/10.1146/annurev-neuro-071013-014030.

Page 139. The brain is usually only 2 percent of the total body weight, but still burns around 20 percent of the calories you ingest: Emma Bryce,

"How Many Calories Can the Brain Burn by Thinking?," Live Science, November 9, 2019, https://www.livescience.com/burn-calories-brain.html/.

Page 139. The relationship between alpha waves and the DMN: Mateusz Rusiniak, Andrzej Wróbel, Katarzyna Cieśla, Agnieszka Pluta, Monika Lewandowska, Joanna Wójcik, Pitor H. Skarżyński, and Tomasz Wolak, "The Relationship Between Alpha Burst Activity and the Default Mode Network," *Acta Neurobiologiae Experimentalis* 78, no. 2 (2018): 92–106.

Page 139. Varying activity in alpha indicates the DMN has different modes: Anthony D. Bowman, Joseph C. Griffis, Kristina M. Visscher, Allan C. Dobbins, Timothy J. Gawne, Mark W. DiFrancesco, and Jerzy P. Szaflarski, "Relationship Between Alpha Rhythm and the Default Mode Network: An EEG-fMRI Study," *Journal of Clinical Neurophysiology* 34, no. 6 (2017): 527–33.

Page 140. A fifth of "aha" moments occur in DMN, and the "aha" experience is perceived as more important than task-oriented "aha" experiences: Shelly L. Gable, Elizabeth A. Hopper, and Jonathan W. Schooler, "When the Muses Strike: Creative Ideas of Physicists and Writers Routinely Occur During Mind Wandering," *Psychological Science* 30, no. 3 (2019): 396–404, https://doi.org/10.1177/0956797618820626.

Page 140. Your problem-solving abilities will be better if you are in DMN: David Rock and Josh Davis, "4 Steps to Having More 'Aha' Moments," *Harvard Business Review*, October 12, 2016, https://hbr.org/2016/10/4-steps-to-having-more-aha-moments/.

Page 140. Do we have seventy thousand thoughts a day? Jon Westenberg, "You Have 70,000 Thoughts Every Single Day—Don't Waste 'Em," *Observer*, May 9, 2017.

Or just four thousand? Creativity professor Joy Bhattacharya works out this figure, as do several other sources. But it's a bit pointless to talk about the number of thoughts, when it's the content that means something to us.

Page 141. When the DMN interacts well with the hippocampus, problem-solving functions better, according to researchers at Oxford: Tim Newman, "The Brain's 'Daydream' Network Is More Active Than We Thought," Medical News Today, October 25, 2017, https://www.medicalnewstoday.com/articles/319846/.

Page 141. Daydreaming and intelligence are related: Christine A. Godwin, Michael A. Hunter, Matthew A. Bezdek, Gregory Lieberman, Seth

Elkin-Frankston, Victoria L. Romero, Katie Witkiewitz, Vincent P. Clark, and Eric H. Schumacher, "Functional Connectivity Within and Between Intrinsic Brain Networks Correlates With Trait Mind Wandering," *Neuropsychologia* 103 (2017): 140.

Page 142. Beth Lapides, *So You Need to Decide*, audible audiobook read by the author (Prince Frederick, MD: Recorded Books, 2021).

Page 144. René Descartes, *Meditations on First Philosophy*, trans. Donald Cress (Indianapolis: Hackett, 1993), 13.

Page 145. Quote from: Kenneth Grahame, *The Wind in the Willows* (London: Methuen, 1908), 4.

Page 145. Mahler rowing and getting the idea for a movement in one of his symphonies, told in: William B. Irvine, *Aha! The Moments of Insight That Shape Our World* (New York: Oxford University Press, 2015), 3.

Page 145. BlueHealth shows that water is extra relaxing: Elle Hunt, "Blue Spaces: Why Time Spent Near Water Is the Secret of Happiness," *Guardian*, November 3, 2019, https://www.theguardian.com/lifeandstyle/2019/nov/03/blue-space-living-near-water-good-secret-of-happiness/.

Page 145. George MacKerron and Susana Mourato, "Happiness Is Greater in Natural Environments," *Global Environmental Change* 23, no. 5 (2013): 992–1000.

Page 146. Depression causes memory and visions of the future to become more general: Hilde Østby and Ylva Østby, *Adventures in Memory: The Science and Secrets of Remembering and Forgetting* (Vancouver: Greystone Books, 2018), 216–18.

Page 146. Depressed artists are less productive: Alice W. Flaherty, *The Midnight Disease: The Drive to Write, Writer's Block, and the Creative Brain* (New York: Houghton Mifflin, 2004), 32.

Page 147. City parks are good for mental health: Victoria Houlden, João Porto de Albuquerque, Scott Weich, and Stephen Jarvis, "A Spatial Analysis of Proximate Greenspace and Mental Wellbeing in London," *Applied Geography* 109 (2019): 102036.

Page 147. Neurologist Oliver Sacks on gardens as medicine and inspiration: Oliver Sacks, "Oliver Sacks: The Healing Power of Gardens," *New York Times*, April 18, 2019, https://www.nytimes.com/2019/04/18/opinion/sunday/oliver-sacks-gardens.html.

Page 147. On how access to nature has an impact on depression, at a population level: Johann Hari, *Lost Connections: Uncovering the Real Causes of*

Depression—and the Unexpected Solutions (London: Bloomsbury, 2018), 123ff.

Page 151. An EEG study carried out on an artist. The connection between meditation and mental pictures, alpha and gamma: Caroline Di Bernardi Luft, Ioanna Zioga, Michael J. Banissy, and Joydeep Bhattacharya, "Spontaneous Visual Imagery During Meditation for Creating Visual Art: An EEG and Brain Stimulation Case Study," *Frontiers in Psychology* 10 (2019), article 210, https://doi.org/10.3389/fpsyg.2019.00210.

Page 152. Eric Klinger's research is quoted in: Manoush Zomorodi, *Bored and Brilliant: How Spacing Out Can Unlock Your Most Productive and Creative Self* (New York: St. Martin's Press, 2017), 22.

Page 153. Raichle's quote is taken from this article (Norwegian only): "Minner om fremtiden og vindskjeve furuer" [Reminiscent of the future and windswept pines], *Dyade*, March 2014, https://dyade.no/tidsskrift/dyade_2014_03_tankestroemmens_tidsalder/minner_om_fremtiden_og_vindskjeve_furuer.

Page 153. Kurt Vonnegut, quoted from: Kurt Vonnegut, *Breakfast of Champions* (New York: Dial Press, 1973), 201.

Page 154. Children quite possibly spend more time in DMN than adults: Manoush Zomorodi, *Bored and Brilliant: How Spacing Out Can Unlock Your Most Productive and Creative Self* (New York: St. Martin's Press, 2017), 22.

Page 154. The development of the DMN in children goes through several stages, and can be influenced by socioeconomic conditions. The development of a number of mental illnesses and ailments may have something to do with challenges in childhood and the maturation of the DMN, according to these researchers: Keila Rebello, Luciana M. Moura, Walter H. L. Pinaya, Luis A. Rohde, and João R. Sato, "Default Mode Network Maturation and Environmental Adversities During Childhood," *Chronic Stress* 2 (2018): 1–10.

Page 154. Sad and happy music and how they affect introspection and the DMN: Liila Taruffi, Corinna Pehrs, Stavros Skouras, and Stefan Koelsch, "Effects of Sad and Happy Music on Mind-Wandering and the Default Mode Network," *Scientific Reports* 7 (2017): 14396.

Page 155. About mild traumatic head injury and the DMN: Chandler Sours, Elijah O. George, Jiachen Zhuo, Steven Roys, and Rao P. Gullapalli, "Hyper-connectivity of the Thalamus During Early Stages Following Mild Traumatic Brain Injury," *Brain Imaging and Behavior* 9, no. 3 (2015): 550–63, https://doi.org/10.1007/s11682-015-9424-2.

Page 155. About micro-changes to the brain after a head injury: Erica J. Wallace, Jane L. Mathias, Lynn Ward, Jurgen Fripp, Stephen Rose, and Kerstin Pannek, "A Fixel-Based Analysis of Micro- and Macro-structural Changes to White Matter Following Adult Traumatic Brain Injury," *Human Brain Mapping* 41, no. 8 (2020): 2187–97.

Page 155. The DMN and depression: Hui-Xia Zhou, Xiao Chen, Yang-Qian Shen, Le Li, Ning-Xuan Chen, Zhi-Chen Zhu, Francisco Xavier Castellanos, and Chao-Gan Yan, "Rumination and the Default Mode Network: Meta-analysis of Brain Imaging Studies and Implications for Depression," *Neuroimage* 206 (2020): 116287, https://doi.org/10.1016/j.neuroimage.2019.116287.

Page 155. Meditation increases the interaction between the DMN and executive function. Disorders in the networks can be associated with both anxiety and depression, and a correlation has been observed with both ADHD and schizophrenia: C. C. C. Bauer, S. Whitfield-Gabrieli, J. L. Diaz, E. H. Pasaye, and F. A. Barrios, "From State-to-Trait Meditation: Reconfiguration of Central Executive and Default Mode Networks," *eNeuro* 6, no. 6 (2019): 0335-18, https://doi.org/10.1523/ENEURO.0335-18.2019.

Page 156. Knitting is very relaxing: Jane E. Brody, "The Health Benefits of Knitting," *New York Times*, January 25, 2016, https://well.blogs.nytimes.com/2016/01/25/the-health-benefits-of-knitting/.

Page 157. On the necessity of isolating oneself: Thure Erik Lund, *Romutvidelser* [Room expansions] (Oslo: Aschehoug, 2019), 67.

Page 158. Max Weber, *The Protestant Ethic and the Spirit of Capitalism*, revised 1920 edition, trans. Stephen Kalberg (New York: Oxford University Press, 2010), 158.

Page 158. Protestants actually work more: Horst Feldmann, "Protestantism, Labor Force Participation, and Employment Across Countries," *American Journal of Economics and Sociology* 66, no. 4 (2007): 795–816.

Page 158. Henrik Ibsen, *Brand*, trans. F. E. Garrett, Everyman's Library (London: J. M. Dent & Sons, 1915), 27, https://archive.org/stream/brand00ibse?ref=ol#page/26/mode/2up.

Page 159. Quote from Rebecca Solnit, *A Field Guide to Getting Lost* (New York: Viking, 2005), 5–6.

Page 159. Quote from Dr. Seuss, *Oh, the Places You'll Go!* (New York: Random House, 1990).

Page 163. Quote from Virginia Woolf, *To the Lighthouse* (New York: Oxford University Press, 1998), 52.

Page 163. Quotes from Siri Hustvedt, *The Shaking Woman or A History of My Nerves* (New York: Henry Holt, 2009), 197.

Page 164. Quote from Lewis Carroll, *Alice's Adventures in Wonderland* (London: Macmillan, 1865), 270.

Page 164. Quote from Siri Hustvedt, *Memories of the Future* (New York: Simon & Schuster, 2019), 348.

Page 168. The median salary for authors in Norway is 120,000 kroner (13,000 US dollars; this article in Norwegian only): Heidi Borud, "Det er blitt vanskeligere å leve av å være forfatter" [It has become more difficult to make a living from being a writer], *Aftenposten*, October 16, 2019, https://www.aftenposten.no/kultur/i/4qvQzo/det-er-blitt-vanskeligere-aa-leve-av-aa-vaere-forfatter?spid_rel=2.

Page 168. Quote from Lewis Carroll, *Through the Looking-Glass, and What Alice Found There* (London: Macmillan, 1871), 228.

Page 170. Quote from Agnes Ravatn in: Hans Petter Blad and Kristine Isaksen, *Min metode: Om sakprosaskriving* [My method: About nonfiction writing] (Oslo: Cappelen Damm, 2019), 46.

Page 170. Manoush Zomorodi, *Bored and Brilliant: How Spacing Out Can Unlock Your Most Productive and Creative Self* (New York: St. Martin's Press, 2017), 38.

Page 170. How to be free of your cell phone: "Bored . . . and Brilliant? A Challenge to Disconnect From Your Phone," *All Things Considered*, NPR, January 12, 2015, https://www.npr.org/sections/alltechconsidered/2015/01/12/376717870/bored-and-brilliant-a-challenge-to-disconnect-from-yourphone?t=1559233226912.

Page 170. Many of the social media and apps we use are unsurprisingly designed to create addiction among users (Norwegian only): Johan Nordstrøm, "Kritiserer Facebook, Twitter og Snapchat for å manipulere sine brukere til å bli avhengige" [Criticized: Facebook, Twitter, and Snapchat for manipulating their users to become addicted], *E24*, July 4, 2018, https://e24.no/digital/sosiale-medier/kritiserer-facebook-twitter-ogsnapchat-for-aa-manipulere-sine-brukere-til-aa-bli-avhengige/24385675.

Page 170. *The Social Dilemma*, directed by Jeff Orlowski (2020, Netflix).

Page 170. Used correctly, social media can make you happy (Norwegian only): Hilde Hartmann Holsten, "Riktig bruk av sosiale medier kan gjøre deg lykkelig" [Proper use of social media can make you happy], *Forskning*, March 6, 2018, https://forskning.no/internett-media-partner/rigtig-bruk-av-sosialemedier-kan-gjore-deg-lykkel/284600.

Page 170. Research on students, anxiety, and cell phones: Emily G. Lattie, Sarah Ketchen Lipson, and Daniel Eisenberg, "Technology and College Student Mental Health: Challenges and Opportunities," *Frontiers in Psychiatry* 10 (2019), 246.

Page 170. Research on what cell phones do to us in relation to anxiety and loneliness: Jon D. Elhai, Jason C. Levine, Robert D. Dvorak, and Brian J. Hall, "Fear of Missing Out, Need for Touch, Anxiety and Depression Are Related to Problematic Smartphone Use," *Computers in Human Behavior* 63 (2016): 509–16.

Page 171. Fairy tale retold by Dana and Milada Stovícková, *Kinesiske eventyr* [Chinese fairy tales], trans. Odd Bang-Hansen (Oslo: Tiden Forlag, 1973).

Page 172. The Oscar Wilde quote is taken from *The Importance of Being Earnest*, act 1, https://www.shmoop.com/importance-of-being-earnest/act-i-full-text-16.html.

CHAPTER 5: HOW TO LEARN LESS AND LESS

Page 177. About Lewis Carroll, in: Jenny Woolf, *The Mystery of Lewis Carroll: Discovering the Whimsical, Thoughtful, and Sometimes Lonely Man Who Created* Alice in Wonderland (New York: St. Martin's Press, 2010), 128.

Page 180. Sverre Malling, *Many a Blossom Shall Its Leaves Unfold* (Oslo: No Comprendo Press, 2017).

Page 183. Professor Markus Lindholm on researcher education in kindergarten (Norwegian only): Markus Lindholm, "Gi ikke arnehagebarna vernebriller" [Do not give kindergarten children safety goggles], *Aftenposten*, October 21, 2018, https://www. aftenposten.no/meninger/kronikk/i/bKx8Av/Gi-ikke-barnehagebarna-vernebriller-Markus-Lindholm.

Page 184. Professor Markus Lindholm on Steiner schools: Markus Lindholm, "Hundre år med barnet som genuint utviklingsprosjekt" [One hundred years with the child as a genuine development project], *Aftenposten*, September 20, 2019, https://www.aftenposten.no/meninger/kronikk/i/mRpwMq/hundre-aar-med-barnet-som-genuint-utviklingsprosjekt-markus-lindholm.

Page 186. About the creativity crisis: Liane Gabora, "What Creativity Really Is—and Why Schools Need It," *The Conversation*, August 30, 2017, https://theconversation.com/what-creativity-really-is-and-why-schools-need-it-81889/.

Page 188. Espen Schaanning on the school system (Norwegian only): Siw Ellen Jakobsen, "Forsker: Skolen I dag skaper flere tapere" [Researcher:

The school today creates more losers], *Forskning*, August 28, 2018, https://forskning.no/partner-skole-og-utdanning-barn-og-ungdom/forsker-skolen-i-dag-skaper-flere-tapere/1222776?fbclid=IwAR1U3m8xhzMi-sEwvObPTZ3MEUwhMxBQf3WaDt3RAcb8r2SDIKRLY91tVaY.

Page 191. Harriet Bjerrum Nielsen on the school system (Norwegian only): https://www.uniforum.uio.no/nyheter/2018/02/harriet-bjerrum-nielsen%3A---skolesystemet-er-i-ferd.html.

Page 191. Harriet Bjerrum Nielsen, *Skoletid. Jenter og gutter fra 1. til 10. klasse* [Schooltime: Girls and boys from 1st to 10th grade] (Oslo: Universitetsforlaget, 2009).

Page 192. Children have less free time now than they did before: Steven John, "5 Ways Kids Spend Time Differently Today Than in the Past," *Business Insider*, September 11, 2018.

Page 196. Yoga can be effective in school. I didn't write about it here, but it partly confirms what I wrote earlier about the vagus nerve, *parasympathicus*, and how meditation increases interaction between the DMN and executive function: "Yoga in Schools Has 'Profound Impact' on Behaviour," BBC News, March 8, 2019, https://www.bbc.com/news/av/uk-england-norfolk-47489958/.

Page 196. About stress among young people (Norwegian only): "Stress, press og psykiske plager blant unge" [Stress, pressure and mental illness among young people], *Ungdata*, January 23, 2020, https://www.ungdata.no/stress-press-og-psykiske-plager-blant-unge.

Page 196. Usage of antidepressants is rising in Norway: Lisbeth Nilsen, "Langt flere tenåringer bruker antidepressiva" [Far more teens are using antidepressants], *Dagens Medisin*, July 8, 2018, https://www.dagensmedisin.no/artikler/2018/08/07/langt-flere-tenaringer-bruker-antidepressiva/.

And in Britain: "Teenagers' Use of Antidepressants Is Rising With Variations Across Regions and Ethnic Groups," National Institute for Health and Care Research, November 18, 2020, evidence.nihr.ac.uk/alert/teenagers-use-of-antidepressants-is-rising-with-variations-across-regions-and-ethnic-groups/.

And in the US: John Elflein, "Antidepressant Use Among Teenagers in the U.S. From 2015–2019, by Gender," Statista, July 17, 2020, https://www.statista.com/statistics/1133612/antidepressant-use-teenagers-by-gender-us/.

And in Canada: Morgan Bocknek, Robert Cribb, and Liam G. McCoy, "Antidepressant Use Among Youth Is Skyrocketing Across Canada.

Prescribing Doctors Say They Have Little Choice as Teens 'Can't Wait Nine Months' for Therapy," *Toronto Star*, April 26, 2021, www.thestar.com/news/investigations/2021/04/26/more-kids-on-antidepressants-in-canada-than-ever-before-prescribing-doctors-say-they-have-little-choice-as-youth-cant-wait-nine-months-for-therapy.html?rf.

Page 201. Raising children according to the TRICK method: Esther Wojcicki, "I Raised Two CEOs and a Doctor. These Are My Secrets to Parenting Successful Children," *Time*, April 26, 2019, http://time.com/5578064/esther-wojcicki-raise-successful-kids/.

Page 201. Michelle Obama, *Becoming* (New York: Crown, 2018).

CHAPTER 6: THE ART OF PAINTING WHITE ROSES RED

Page 208. The survey conducted by Gallup in 142 countries: "State of the Global Workplace: 2022 Report," Gallup, June 14, 2022, https://www.gallup.com/workplace/349484/state-of-the-global-workplace-2022-report.aspx.

Page 208. In Norway, 135,000 employees have suicidal thoughts at work (Norwegian only): Gunn Evy Auestad, "135.000 blir mobba til sjølvmordstankar" [135,000 are bullied into suicidal thoughts], NRK, October 4, 2015, https://www.nrk.no/norge/135.000-blir-mobba-til-sjolvmordstankar1.12581266.

Page 208. Sexual harassment at work in Norway (Norwegian only): Tiril Mettesdatter Solvang, "Advokat: -Det er ulovlig å ikke melde fra om seksuell trakassering på job" [Lawyer: It is illegal not to report sexual harassment at work], NRK, October 24, 2017, https://www.nrk.no/norge/kollegaer-har-ansvar-for-a-melden-omseksuell-trakassering-1.13747042.

Page 208. Antidepressant use in the USA: Benedict Carey and Robert Gebeloff, "Many People Taking Antidepressants Discover They Cannot Quit," *New York Times*, April 7, 2018, https://www.nytimes.com/2018/04/07/health/antidepressants-withdrawal-prozac-cymbalta.html.

Page 208. Anne Gunn Halvorsen, *Stress og korleis leve med det* [Stress and how to live with it] (Oslo: Samlaget, 2019).

Page 214. Stefan Sagmeister, "The Power of Time Off," October 1, 2009, TED video, 17:24, https://www.ted.com/talks/stefan_sagmeister_the_power_of_time_off.

Tina Essmaker, "Stefan Sagmeister," Great Discontent, June 23, 2014, https://thegreatdiscontent.com/interview/stefan-sagmeister.

Page 215. About the four-day week (Norwegian only): Håvard Hjorthaug Vege, "Flere selskap har testet fire dagers arbeidsuker tils tor suksess"

[Several companies have tested four-day workweeks to great success], *Nettavisen*, September 18, 2018, https://www.nettavisen.no/nyheter/utenriks/flere-selskap-har-testetfire-dagers-arbeidsuker-til-stor-suksess/3423537876.html.

Page 215. Dennis Nørmark and Anders Fogh Jensen, *Pseudowork: How We Ended Up Being Busy Doing Nothing*, trans. Tam McTurk (Copenhagen: Gyldendal, 2021).

Page 216. Miriam Lund Knapstad, "Det er liksom fint å ha det travelt . . ." [It's kind of nice to be busy . . .], *Aftenposten*, June 8, 2018, https://www.aftenposten.no/amagasinet/i/kazLgX/det-er-liksom-fint-aa-ha-det-travelt-men-det-burde-vaert-finere-aa-kunne-si-jeg-er-ferdig-med-det-jeg-skal-gjoere-i-dag-saa-jeg-gaar-klokken-ett.

Page 216. Mike Pearl, "Burnout Is Real, and the Solution Is Radically Changing How We Work," *Vice*, May 30, 2019, https://www.vice.com/en_us/article/ywy5eg/burnout-is-real-and-the-solution-is-radically-changing-how-we-work.

Page 218. Quote from Bertrand Russell, *In Praise of Idleness* (Milton Park, UK: Routledge Classics, 2004), 6–7.

Page 219. An op-ed in a Norwegian newspaper about the standardization of psychiatric treatment (Norwegian only): Sondre Risholm Liverød, "Pakkeforlop og bullshit" [Package process and bullshit], *Dagsavisen*, May 27, 2019, https://www.dagsavisen.no/debatt/pakkeforlop-og-bullshit-1.1529670.

Page 222. How to repair your clothes (Norwegian only): Knut-Erik Helle, "Gjør miljømarerittet I klesskapet grønnere" [Make the environmental nightmare in the wardrobe greener], *Framtiden*, September 20, 2017, https://www.framtiden.no/gronne-tips/klar/gjor-miljomarerittet-i-klesskapet-gronnere.html.

Page 222. And how to repair broken porcelain: Natalie Parkin, "Parkin: Chinese Kintsugi Teaches That Broken Is Beautiful," *Daily Utah Chronicle*, March 5, 2018, https://dailyutahchronicle.com/2018/03/05/parkin-chinese-kintsugi-teaches-broken-beautiful/.

Page 223. Siw Aduvill, *Hvile: Alt du vinner ved å la være* [Rest: All that you gain by letting go] (Oslo: Tiden Forlag, 2019).

CHAPTER 7: THE WALRUS AND THE CARPENTER

Page 229. Emily W. Sunstein, *Mary Shelley: Romance and Reality* (Baltimore: Johns Hopkins University Press, 1991).

Page 229. Mary Shelley, *Frankenstein; Or, The Modern Prometheus* (New York: Penguin Classics, 1992).

Page 229. About the Year Without a Summer, and those who sheltered at Villa Diodati (incidentally, the doctor's [John William Polidori] story also became quite famous in retrospect; his story about a vampire inspired Bram Stoker's *Dracula*): Greg Buzwell, "Mary Shelley, *Frankenstein* and the Villa Diodati," British Library, May 15, 2014, https://www.bl.uk/romantics-and-victorians/articles/mary-shelley-frankenstein-and-the-villa-diodati.

Page 230. Ada Lovelace and the invention of programming: "What Did Ada Lovelace's Program Actually Do?," *Two-Bit History* (blog), August 18, 2018, https://twobithistory.org/2018/08/18/ada-lovelace-note-g.html. Stephen Wolfram, "Untangling the Tale of Ada Lovelace," *Writings* (blog), December 10, 2015, https://writings.stephenwolfram.com/2015/12/untangling-the-tale-of-ada-lovelace/.

Page 231. Quote by Ada Lovelace: "King, Ada, Countess of Lovelace (1815–1852) a.k.a. Ada Byron," English Heritage, https://www.english-heritage.org.uk/visit/blue-plaques/ada-byron/.

Page 231. Claire L. Evans on Ada Lovelace in *Broad Band: The Untold Story of the Women Who Made the Internet* (New York: Penguin, 2018), 13–22.

Page 234. Marcus du Sautoy, *The Creativity Code: Art and Innovation in the Age of AI* (Cambridge, MA: Harvard University Press, 2019), 44ff.

Page 237. Per Kristian Bjørkeng, *Kunstig intelligens: Den usynlige revolusjonen* [Artificial intelligence: The invisible revolution] (Oslo: Vega, 2018).

Page 240. Bår Stenvik on Zapffe and "The Last Messiah" (Norwegian only): Audun Lindholm and Bår Stenvik, "Mennesket: problem eller løsning?" [Man: problem or solution?], *Vagant*, May 14, 2012, http://www.vagant.no/mennesket-problem-eller-losning/.

Page 245. Siri Dokken's bronze-winning submission to the World Humor Awards: "Pris til Siri Dokken i World Humor Awards" [Award for Siri Dokken at the World Humor Awards], Grafill, September 6, 2019, https://www.grafill.no/nyheter/pris-til-siri-dokken-i-world-humor-awards.

Page 246. Thure Erik Lund, *Romutvidelser* [Room expansions] (Oslo: Aschehoug, 2019).

Page 248. About Mary Shelley and her children: Suzanne Burdon, "Mary Shelley and Motherhood," author blog, May 13, 2017, http://www.suzanneburdon.com/blog/2017/5/13/mary-shelley-and-motherhood.

CHAPTER 8: I FIND ALICE

Page 252. Simon Critchley, *Very Little... Almost Nothing: Death, Philosophy, Literature* (Milton Park, UK: Routledge, 1997), 29.

Page 253. Simon Critchley, *Bowie* (New York: OR Books, 2016).

Page 253. Marte Spurkland, *Pappas runer* [Dad's runes] (Oslo: Cappelen Damm, 2019).

Page 256. Else Kåss Furuseth, *Else går til psykolog* [Else goes to the psychologist] (Oslo: Cappelen Damm, 2018).

Page 257. Quotes from: William Shakespeare, *Hamlet* (New York: Simon & Schuster, 1992).

Page 258. Siri Hustvedt, *The Shaking Woman or A History of My Nerves* (New York: Henry Holt, 2009), 3.

Page 258. Haruki Murakami on how voids are the starting point for his writing, in an interview in the *New Yorker*: Deborah Treisman, "Haruki Murakami on How Memory Can Trigger a Story," *New Yorker*, February 10, 2020, https://www.newyorker.com/books/this-week-in-fiction/haruki-murakami-02-17-20/amp.

Page 258. Alice W. Flaherty, *The Midnight Disease: The Drive to Write, Writer's Block, and the Creative Brain* (New York: Houghton Mifflin, 2004), 11.

Page 259. Alice W. Flaherty: *The Midnight Disease: The Drive to Write, Writer's Block, and the Creative Brain* (New York: Houghton Mifflin, 2004), 43.

Page 262. Vera Micaelsen's final message to her young readers, in (Norwegian only): Wenche Fuglehaug Falls, "Hun skrev en siste hilsen til de yngste. Den rører tusenvis." [She wrote a final farewell to her youngest readers. Thousands are touched.] *Aftenposten*, November 10, 2018, https://www.aftenposten.no/norge/i/LORpw1/hun-skrev-en-siste-hilsen-til-de-yngste-den-roerer-tusenvis.

CHAPTER 8½: THE DODO'S LAMENT

Page 266. One million species are in danger of extinction: "1 Million Species at Risk of Extinction, Intergovernmental Report Finds," *Yale Environment 360*, May 6, 2019.
United Nations, "UN Report: Nature's Dangerous Decline 'Unprecedented'; Species Extinction Rates 'Accelerating,'" *Sustainable Development Goals* (blog), May 6, 2019, https://www.un.org/sustainabledevelopment/blog/2019/05/nature-decline-unprecedented-report/.

Page 266. Jenny Woolf, *The Mystery of Lewis Carroll: Discovering the Whimsical, Thoughtful, and Sometimes Lonely Man Who Created* Alice in Wonderland (New York: St. Martin's Press, 2010).

Page 266. Lewis Carroll and "the Oxford Dodo": Annetta Black, "Alice, Oxford, and the Dodo," *Atlas Obscura*, November 19, 2013, https://www.atlasobscura.com/articles/alice-oxford-and-the-dodo.

Page 266. Dodgson and the Dodo: Laura Geggel, "5 Odd Facts About Lewis Carroll," Live Science, July 3, 2015, https://www.livescience.com/51438-lewis-carroll-wonderland.html.

Page 266. Many of the animals from Lewis Carroll's universe are now threatened with extinction:
Walruses: https://www.endangered.org/animal/pacific-walrus/
Flamingos: https://www.iucnredlist.org/species/22697360/131878173
Hedgehogs: https://ptes.org/campaigns/hedgehogs/
Oysters: https://www.independent.co.uk/environment/nature/wild-oysters-indanger-of-extinction-2205743.html

Page 267. Chris Field in: Hilde Østby and Ylva Østby, *Adventures in Memory: The Science and Secrets of Remembering and Forgetting* (Vancouver: Greystone Books, 2018), 262, 264.

Page 267. Per Espen Stoknes, *Det vi tenker på når vi prøver å ikke tenke på global oppvarming* [What we think of when we try not to think about global warming] (Oslo: Tiden Forlag, 2017).

Page 267. From apocalypse to action: Per Espen Stoknes, "How to Transform Apocalypse Fatigue Into Action on Global Warming," November 17, 2017, TED video, 14:50, https://www.ted.com/talks/per_espen_stoknes_how_to_transform_apocalypse_fatigue_into_action_on_global_warming.

Page 268. Quote from Thure Erik Lund, *Romutvidelser* [Room expansions] (Oslo: Aschehoug, 2019), 187.

Page 270. About food and climate: "The EAT-Lancet Commission on Food, Planet, Health," EAT Forum, https://eatforum.org/eat-lancet-commission/.

Page 270. About the Impossible Burger: Juliet Sear, "What Does the Impossible Burger Taste Like?," *BBC Good Food*, https://www.bbcgoodfood.com/howto/guide/what-does-impossible-burger-taste.

Page 270. What is the carbon footprint of our food?: Nassos Stylianou, Clara Guibourg, and Helen Briggs, "Climate Change Food Calculator: What's Your Diet's Carbon Footprint?," BBC, August 9, 2019, https://www.bbc.com/news/science-environment-46459714.

Page 271. Refugee tent that can collect rainwater: "Inspiring Woman Invents Refugee Tents That Collect Rainwater and Store Solar Energy," Egyptian Streets, December 27, 2018, https://egyptianstreets

.com/2018/12/27/female-architect-invents-refugee-tents-that-collect-rainwater-and-store-solar-energy/?fbclid=IwAR3mgnOrN1ylSELj7hvvUu2xb71HTnhpEyKK885.

Page 271. Bill Gates works with a kind of "wall" that draws CO_2 out of the atmosphere: John Vidal, "How Bill Gates Aims to Clean Up the Planet," *Guardian*, February 4, 2018, https://www.theguardian.com/environment/2018/feb/04/carbon-emissions-negative-emissions-technologies-capture-storage-bill-gates.

Page 271. A fast-growing coral that can save coral reefs: Evan Fleischer, "Scientist's Accidental Discovery Makes Coral Grow 40× Faster," Big Think, December 5, 2018, https://bigthink.com/surprising-science/fast-growing-coral-discovery-could-revitalize-oceans.

Page 271. About the film *2040* in the *Guardian*: Damon Gameau, "A Vision of 2040: Everything We Need for a Sustainable World Already Exists," *Guardian*, May 20, 2019, https://www.theguardian.com/environment/2019/may/21/determination-and-passion-how-these-renewable-energy-resources-can-save-our-planet.

Page 274. I have also read: Al Gore, *An Inconvenient Sequel: Truth to Power* (New York: Crown, 2017); Jonathan Safran Foer, *We Are the Weather* (New York: Farrar, Straus and Giroux, 2019); and Dag O. Hessen, *Verden på vippepunktet* [The world at the tipping point] (Oslo: Res Publica, 2020).

Page 275. About the Future Library: https://www.futurelibrary.no/.

Page 279. Head injuries and work situations. This article addresses permanent brain injuries, but temporary ones can also lead to changes in a person's private and work situation (Norwegian only): Anita Puhr, Bård Fossum, Børre Hansen, and Anne Tove Thorsen, "Tilbakeføring til arbeid etter ervervet hjerneskade" [Return to work after an acquired brain injury], *Psykologi*, August 5, 2011, https://psykologtidsskriftet.no/fagartikkel/2011/08/tilbakeforing-til-arbeid-etter-ervervet-hjerneskade.

Page 280. William Shakespeare, *The Tempest* (London: The Arden Shakespeare, 1954).

If you have suffered a head injury, consult a doctor. Whatever you do, avoid reading, and that goes for this book.